职业技术教育课程改革规划教材
光机电专业国家级教学资源库系列教材

光电技术应用专业英语

GUANG DIAN JISHU YINGYONG ZHUANYE YINGYU

主　编　王凌波　王九程
副主编　周　利　吴天慧　王　琛
参　编　张　蕾　颜红梅　宋露露　刘晓竹
主　审　吴晓红

中国·武汉

内 容 简 介

本书是以光电技术行业对复合型人才的需求为导向编写的专业英语教材,主要涉及电学、光学、激光基础、光电探测技术、激光加工技术和光纤通信等方面的内容。全书共 6 个单元,每个单元均包括 4 篇课文、2 篇拓展阅读材料、听说训练、翻译训练,以及任务单。每篇课文都配有词汇注释及练习。

本书体系合理,题材广泛,内容丰富,可作为高等职业院校、高等专科院校、成人高校、民办高校、本科院校的二级职业技术学院和中等职业院校光电技术应用专业的教学用书,也可作为相关工程技术人员和社会从业人士的参考书和培训用书。

图书在版编目(CIP)数据

光电技术应用专业英语/王凌波,王九程主编. —武汉:华中科技大学出版社,2018.9
职业技术教育课程改革规划教材. 光机电专业国家级教学资源库系列教材
ISBN 978-7-5680-4582-7

Ⅰ.①光… Ⅱ.①王… ②王… Ⅲ.①光电技术-英语-职业教育-教材 Ⅳ.①TN2

中国版本图书馆 CIP 数据核字(2018)第 207327 号

光电技术应用专业英语　　　　　　　　　　　　　　　　　王凌波　王九程　主编
Guangdian Jishu Yingyong Zhuanye Yingyu

策划编辑:王红梅	
责任编辑:刘　平	
封面设计:秦　茹	
责任校对:李　弋	
责任监印:赵　月	
出版发行:华中科技大学出版社(中国·武汉)	电话:(027)81321913
武汉市东湖新技术开发区华工科技园	邮编:430223
录　　排:武汉楚海文化传播有限公司	
印　　刷:武汉科源印刷设计有限公司	
开　　本:787mm×1092mm　1/16	
印　　张:14.25	
字　　数:393 千字	
版　　次:2018 年 9 月第 1 版第 1 次印刷	
定　　价:36.80 元	

本书若有印装质量问题,请向出版社营销中心调换
全国免费服务热线:400-6679-118　竭诚为您服务
版权所有　侵权必究

序

  作为新兴的行业、产业,我国光电技术的发展一日千里,光电产业对我国经济社会的巨大作用日益凸显。我国光电与激光市场十几年来始终保持两位数的高速增长,2010年我国光电与激光产业的市场规模已经突破千亿。随着信息技术、激光加工技术、激光医疗与光子生物学、激光全息、光电传感、显示技术及太阳能利用等技术的快速发展,我国光电与激光产业市场规模将进一步扩大。

  随着光电产业的不断产展,对光电技术人才的需求越来越大,高等职业院校光电技术方面的专业建设也越来越受到重视。作为其中的重要部分,光电专业教材建设目前虽然取得了一定的成果,但还无法满足产业发展对人才培养的需求,尤其是面向职业教育的专业教材更是屈指可数,很多学校都只能使用自编的校本教材。值此国家"十二五"规划实行之际,编写和出版职业院校使用的光电专业教材既迫在眉睫,又意义重大。

  华中科技大学出版社充分依托"武汉·中国光谷"的区域优势,在相继开发分别面向全国"211"重点大学和普通本科大学光电专业教材的基础上,又倾力打造了这套面向全国职业院校的光电技术专业系列教材。在组织过程中,华中科技大学出版社邀请了全国所有开设有光电专业的职业院校的专家、学者,同时与国内知名的光电企业合作,在国家光电专业教指委专家的指导下,齐心协力,求同存异,取长补短,共同编写了这套应用范围最广的光电专业系列教材。参与本套教材建设的院校大多是国家示范院校或国家骨干院校,他们在光电专业建设上取得了良好的成绩。参与本套教材编写的教师,基本上是相关国家示范院校或国家骨干院校光电专业的带头人和长期在一线教学的教师,非常了解光电专业职业教育的发展现状,具有丰富的教学经验,在全国光电专业职业教育领域中也有着广泛的影响力。此外,本套教材还吸收了大量有丰富实践经验的企业高级工程技术人员参与编写,参考了企业技术创新资料,把教学和生产实践有效地结合在一起。

  本套教材的编写基本符合当前教育部对职业教育改革规划的精神和要求,在坚持工作过程系统化的基础上,重点突出职业院校学生职业竞争力的培养和锻炼,以光电行业对人才需求的标准为原则,密切联系企业生产的实际需求,对当前的光电专业职业教育具有很好的指导作用。

  本套教材具有以下鲜明的特点。

  课程齐全。本套教材基本上包括了光电专业职业教育的专业基础课和光电子、光器件、光学加工、激光加工、光纤制造与通信等各个领域的主要专业课，门类齐全，是对光电专业职业教育一次有效、有益的整理与总结。

内容新颖。 本套教材密切联系当前光电技术的最新发展动态，在介绍基本原理、知识的基础上，注重吸收光电专业的新技术、新理念、新知识，并重点介绍了它们在生产实践中的应用，如《平板显示器的制造与测试》《LED封装与测试》。

针对性强。 本套教材结合职业教育和职业院校的实际教学现状，非常注重知识"可用、够用、实用"，如《工程光学基础教程》《激光原理与技术》。

原创性强。 本套教材是在相关国家示范院校或国家骨干院校长期使用的自编校本教材的基础上形成的，既经过了教学实践的检验，又进行了总结、提高和创新，如《光纤光学基础》《光电检测技术》。其中的一些教材，在光电专业职业教育中更是首创，如《光电子技术英语》《光学加工工艺》。

实践性强。 本套教材非常注重实验、实践、实训"易实施、可操作、能拓展"。书中不少实验、实训基本上都是企业实践中的生产任务，有的甚至是整套生产线上任务的实施，如《激光加工设备与工艺》《光有源无源器件制造》。

我十分高兴能为本套教材写序，并乐意向各位读者推荐，相信读者在阅读这套教材后会和我一样获得深刻印象。同时，我十分有幸认识本套教材的很多主编，如武汉职业技术学院的吴晓红、武汉软件工程职业学院的王中林、南京信息职业技术学院的金鸿、苏州工业园区职业技术学院的吴文明、福建信息职业技术学院的林火养等老师，知道他们在光电专业职业教育中的造诣、成绩及影响；也和华中科技大学出版社有过合作，了解他们在工科教材出版，尤其是在光电技术（信息）方面教材出版上的成绩和成效。我相信，由他们一起编写、出版本套教材，一定会相得益彰。

本套教材不仅能用于指导当前光电专业职业教育的教学，也可用于指导光电行业企业员工培训或社会职业教育的培训。

中国光学学会激光加工专业委员会主任

2011年8月24日

前 言

专业英语是通过阅读英文专业文献来获取专业信息的一门课程，所涉及的专业文献一般是学生已经了解和掌握的知识。从20世纪60年代初起，人们在有关英语教学的文献上就经常可以看到"特殊目的英语"(ESP：English for Specific Purposes)这样的字眼，它意指英语教学是为了一个特殊的、实用的目的而进行的。

随着社会和经济的发展，人们对人才素质的要求也越来越高，这给现代教育带来了巨大的挑战。21世纪要求的人才应该具有综合素质好、专业精通、外语基础扎实、听说读写综合技能过硬、适应性强等特色。

2011年，我们开发了《光电子技术专业英语》教材，得到了广大师生的认可。随着教育信息化建设的推进，职业教育改革不断深入，2018年我们推出了《光电技术应用专业英语》，其鲜明特色是"职业性"，即为光电专业和相关专业学生提供将来从事其行业必要的英语能力支撑（包括英文科技资料的阅读能力、一般语言交流能力等），直接或间接促进学生的专业学习。教材涉及电学、光学、激光基础、光电探测技术、激光加工技术和光纤通信这六个方面的内容。

全书共六个单元，每个单元均包括专业技术性文章和拓展阅读，以及听说训练、翻译训练、单元任务单。

1) READING IN DEPTH 与 EXTENSIVE READING 前后呼应，旨在逐步提高学生的专业英语阅读能力；READING IN DEPTH 部分每篇文章的开始均配有专业知识目标。

2) CAREER SPEAKING 和 WRITING 两部分对应的是职场能力，内容紧扣学生的求职就业、职业发展需求，凸显语言学习的实用性。

3) TRANSLATION 注重将翻译技巧浅显化，便于学生掌握、运用。

4) UNIQUE ENGLISH 和 Warm-Up Activity 尝试提高专业英语教学的趣味性。

5) TASK 部分，引导学生根据该单元的知识，表达相关领域的专业信息与见解，并制作微视频。近年来互联网技术飞速发展，智能终端的普及与应用使得这种图、文、声的表达与交流方式能够流畅地在课堂教学中实现。

为了方便教学，各单元课文及阅读材料均标注生词和新短语。另外，本书还配有电子教案和练习题参考答案，向采纳本书作为教材的教师免费提供。

其中，READING IN DEPTH、EXTENSIVE READING 的编写分工如下：UNIT 1 由武汉职业技术学院王凌波编写，UNIT 2 由王凌波、武汉大学周利编写，UNIT 3 由武汉软件工程职业学院王九程编写，UNIT 4 由王凌波、汉口学院王琛、湖北大学张蕾编写，UNIT 5 由武汉软件工程职业学院颜红梅、吴天慧编写，UNIT 6 由王琛、武汉市仪表电子学校刘晓竹、武汉职业技术学院宋露露编写；CAREER SPEAKING、WRITING、TRANSLATION 由武汉软件工程职业学院吴天慧编写；TASK 由王凌波、张蕾、王琛开发与编写；UNIQUE ENGLISH 由王九程编写。本书由王凌波、王九程共同完成设计、统稿和编排，主审是武汉职业技术学院吴晓红教授。

由于编者水平有限，疏漏和不妥之处在所难免，恳请读者不吝指正。

反馈意见、免费索取电子教案及参考答案请发邮件至邮箱 wlb0125@163.com。

编　者

2018 年 5 月

CONTENTS

Unit 1	Basic Circuit and Its Applications	1
	READING IN DEPTH	2
	1.1 Basic Circuit Concepts	2
	1.2 Capacitance and Inductance	6
	1.3 Voltage Dividers	10
	1.4 Filters	14
	CAREER SPEAKING	17
	Job Interview	17
	WRITING	21
	Resume	21
	EXTENSIVE READING	23
	1.5 Ohm's Law	23
	1.6 What Is Electricity	24
	TRANSLATION	25
	Features of Scientific English	25
	TASK	27
	Basic Circuit Concepts	27
	UNIQUE ENGLISH	30
	Brain Teaser	30
Unit 2	Optics	31
	READING IN DEPTH	32
	2.1 Nature of Light	32
	2.2 Reflection and Refraction	36
	2.3 Numerical Aperture (NA)	41
	2.4 Electromagnetic Radiation	44

CAREER SPEAKING ... 48
Meeting Guests ... 48
WRITING ... 53
Letter of Thanks ... 53
EXTENSIVE READING ... 55
2.5 What Is Light ... 55
2.6 Convex Lens ... 55
TRANSLATION ... 56
Translation of Words and Phrases ... 56
TASK ... 58
Electromagnetic Radiation ... 58
UNIQUE ENGLISH ... 62
Incredible Sentences ... 62

Unit 3 Laser ... 65
READING IN DEPTH ... 66
3.1 This Is a Laser ... 66
3.2 Applications of Lasers ... 71
3.3 Laser Construction ... 77
3.4 Internet and Lasers ... 82
CAREER SPEAKING ... 88
Introducing Enterprises ... 88
WRITING ... 93
Letter of Congratulation ... 93
EXTENSIVE READING ... 94
3.5 A Prospectus about Laser Machines ... 94
3.6 Using Lasers ... 96
TRANSLATION ... 98
Translation of Noun Clauses ... 98
TASK ... 99
Laser Construction ... 99
UNIQUE ENGLISH ... 104
Funny Chinglish ... 104

| Unit 4 | **Optical Detection** | 105 |

READING IN DEPTH ... 106
    4.1 Photoelectric Effect ... 106
    4.2 Photodiode ... 111
    4.3 Optical Detection Techniques ... 115
    4.4 CCD ... 120

CAREER SPEAKING ... 124
    Making Calls ... 124

WRITING ... 129
    Letter of Apology ... 129

EXTENSIVE READING ... 131
    4.5 LED Technology ... 131
    4.6 Optical Storage Technology ... 132

TRANSLATION ... 133
    Translation of Attributive Clauses ... 133

TASK ... 134
    Optical Detection ... 134

UNIQUE ENGLISH ... 140
    Exotic Sentences ... 140

Unit 5 Laser Processing ... 141

READING IN DEPTH ... 142
    5.1 Laser Cutting ... 142
    5.2 Laser Marking ... 147
    5.3 Laser Welding ... 153
    5.4 Laser Perforating ... 159

CAREER SPEAKING ... 164
    After-Sales Service ... 164

WRITING ... 168
    E-mail ... 168

EXTENSIVE READING ... 169
    5.5 Everyday Use of Lasers ... 169
    5.6 Laser Safety ... 171

TRANSLATION · · · · · · 173
How to Translate with Aid of the Internet and Software · · · · · · 173
TASK · · · · · · 175
Laser Marking · · · · · · 175
UNIQUE ENGLISH · · · · · · 181
Funny Proverbs · · · · · · 181

Unit 6 Fiber-Optic Communication · · · · · · 183
READING IN DEPTH · · · · · · 184
6.1 Brief Introduction of Fiber-Optic Communication · · · · · · 184
6.2 Typical Optical Transmitter—LEDs · · · · · · 189
6.3 Typical Optical Receivers · · · · · · 194
6.4 Comparison with Electrical Transmission · · · · · · 199
CAREER SPEAKING · · · · · · 204
Business Meeting · · · · · · 204
WRITING · · · · · · 209
Memo · · · · · · 209
EXTENSIVE READING · · · · · · 210
6.5 Optical Fibers · · · · · · 210
6.6 Fiber to the Home · · · · · · 212
TRANSLATION · · · · · · 213
Translation of Adverbial Clauses · · · · · · 213
TASK · · · · · · 215
Fiber-Optic Communication · · · · · · 215
UNIQUE ENGLISH · · · · · · 215
Mother · · · · · · 215

References · · · · · · 217

Unit 1

Basic Circuit and Its Applications

In this unit, you are going to learn the following contents:

READING IN DEPTH

1.1 Basic Circuit Concepts

1.2 Capacitance and Inductance

1.3 Voltage Dividers

1.4 Filters

CAREER SPEAKING

Job Interview

WRITING

Resume

EXTENSIVE READING

1.5 Ohm's Law

1.6 What Is Electricity

TRANSLATION

Features of Scientific English

TASK

Basic Circuit Concepts

UNIQUE ENGLISH

Brain Teaser

READING IN DEPTH

1.1 Basic Circuit Concepts

Learning objectives:

In this section, you will:
- understand the meaning of direct current (DC).
- learn about voltage, current, and resistance.
- master the relationship between voltage, current, and resistance in an electrical circuit.

Warm-Up Activity

Do you know the names of the components below?

Picture 1

Picture 2

Picture 3

Picture 4

Picture 5

Text

Fig. 1.1 shows the basic type of an electrical **circuit** in the form of a **block diagram**. It consists of a source of electrical energy, some sort of **load** to make use of that energy, and electrical conductors connecting the source and the load.

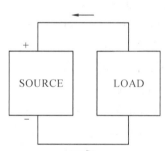

Fig. 1.1 A basic DC circuit

The electrical source has two **terminals**, **designated** positive (+) and negative (−). **As long as** there is an unbroken connection from source to load and back again as shown here, **electrons** will be pushed from the negative terminal of the source, through the load, and then back to the positive terminal of the source. The arrows show the direction that electron current flows through this circuit. Because the electrons are always moving in the same direction through the circuit, their **motion** is known as a direct current (DC).

The source can be any source of electrical energy. In practice, there are three general possibilities: it can be a battery, an electrical generator, or some sort of electronic power supply.

The load is any device or circuit powered by **electricity**. It can be as simple as a light bulb or as complex as a modern high-speed computer.

circuit n. 电路
block diagram 结构图,
方块图,简图
load n. 负载

terminal n. 终端,接线端
designate v. 指定,指派
as long as
只要
electron n. 电子

motion n. 运动,动作

electricity n. 电能,电

The electricity provided by the source has two basic characteristics, called **voltage** and **current**. These are **defined** as follows.

Voltage

The electrical "pressure" that causes free electrons to travel through an electrical circuit, also known as **electromotive force** (**emf**). It is measured in volts.

Current

The amount of electrical charge (the number of free electrons) moving past a given point in an electrical circuit per unit of time. Current is measured in amperes.

The load, in turn, has a characteristic called resistance. By definition:

Resistance

That characteristic of a medium which opposes the flow of electrical current through itself. Resistance is measured in ohms.

The relationship between voltage, current, and resistance in an electrical circuit is fundamental to the operation of any circuit or device. **Verbally**, the amount of current flowing through a circuit is directly proportional to the applied voltage and inversely proportional to the circuit resistance. By explicit definition, one volt of electrical pressure can push one ampere of current through one ohm of resistance. Two volts can either push one ampere through a resistance of two ohms, or push two amperes through a resistance of one ohm. Mathematically,

$$E = IR$$

where E = the applied voltage, or EMF;

I = the circuit current;

R = the resistance in the circuit.

voltage *n.* 电压
current *n.* 电流
define *vt.* 定义

electromotive force 电动势

verbally *adv.* 用文字来表达

Get to Work

Ⅰ. Matching

Directions: Match the two columns.

1. current   A. electromotive force
2. amp   B. the flow of electrons
3. voltage   C. the unit of resistance
4. ohm   D. the unit in which current is measured
5. DC   E. the direction of current is constant

II. Understanding Checking

Directions: Mark Y (for YES) if the statement agrees with the information given in the text; N (for NO) if the statement contradicts the information given in the text; NG (for NOT GIVEN) if the information is not given in the text.

( ) 1. One kilovolt equals one thousand volts.

( ) 2. The flow of electrons through a conductor is called resistance.

( ) 3. Current flow is represented by the letter symbol *I*.

( ) 4. The opposition to current is called electrical resistance.

( ) 5. The relationship between voltage, current, and resistance in an electrical circuit is fundamental to the operation of any circuit or device.

III. Passage Completion

Directions: Fill in each of the blanks with one of the words or expressions in the box, making changes if necessary.

| constant | cycle | change | negative | repeat |

To have a better understanding of alternating current and voltage, it is desirable to begin with a consideration of the general situation. A function of time is called alternating when after a time interval or period, it repeats a previous succession of positive and __1__ values. In other words, such an action __2__ itself exactly over equal intervals of time. So an alternating current is a rate of electricity which does not have a __3__ value in time but grows to a maximum value, decreases, __4__ its direction reaches a maximum value in the new direction, returns to its original value and then repeats this __5__ at an indefinite number of times.

IV. Translation

Directions: Complete the sentences by translating into English according to the Chinese given in brackets.

1. Modern advances in the fields of computer, control system communication _____ _____ (与电子工业密切相关).

2. When it is desirable to express a magnitude of current _____ (比安培小), the milliampere and microampere units are used.

3. _____ (电流、电压和电阻的关系) in an electrical circuit is fundamental to the operation of any circuit or device.

4. There are _____ (大量的符号) which represent an equally large range of electronic components.

5. The _____ (电流的单位) is the ampere.

1.2 Capacitance and Inductance

Learning objectives:

In this section, you will:
- ● understand the charging process of a capacitor.
- ● learn about three important factors of a capacitor.
- ● comprehend the use of capacitors in a radio receiver.
- ● learn how to make an inductor by hand.
- ● calculate how much energy an inductor can store.

Warm-Up Activity

Do you know the names of the components below?

Picture 1

Picture 2

Picture 3

Picture 4

Picture 5

Text

Capacitance

Electrical energy can be stored in an **electric** field. The **device** capable of doing this is called a **capacitor** or a **condenser**.

A simple capacitor consists of two metallic plates separated by a **dielectric**. If a capacitor is connected to a battery, the electrons will flow out of the negative terminal of the battery and accumulate on the condenser plate connected to that side. At the same time, the electrons will leave the plate connected to the positive terminal and flow into the battery to make the potential difference just the same as that of the battery. Thus the capacitor is said to be **charged**.

The capacitance is directly **proportional** to the dielectric constant of the material and to the area of the plates and inversely to the distance of the plates. It is measured in **farads**. When a change of one volt per second across it causes the current of one ampere to flow, the condenser is said to have the capacitance of one farad. However, farad is too large a unit to be used in radio calculation, so microfarad (one millionth of a farad) and the picofarad (10^{-12} farad) are generally used.

The amount of the stored energy of a charged condenser is proportional to the applied voltage and its capacitance. The capacitance of a condenser is determined by three important factors, namely the area of the plate surface, the space between them and dielectric material. The larger the plate area, the smaller the space between plates, the greater the capacitance.

One of the condensers used in radio receiver is a variable condenser, whose capacitance can be varied by turning the plates. It is used in the receiver for turning and varying capacitance in the circuit so as to pick up the desired signals of different wavelengths.

Inductance

Any current will create a **magnetic** field, so in fact every current carrying wire in a circuit acts as an inductor! However, this type of **stray** inductance is typically **negligible**, just as we can usually ignore the stray resistance of our wires and only take into account the actual resistors. To store any appreciable amount of magnetic energy, one

---

electrical *adj.* 电的,有关电的
electric *adj.* 电动的,电气的,以电为动力的
device *n.* 装置,设备
capacitor *n.* 电容器
condenser *n.* 电容器
dielectric *n.* 电介质,绝缘体
charge *n.* 负荷,电荷 *v.* 充电
proportional *adj.* 比例的,成比例的
farad *n.* [电]法拉(电容单位)

magnetic *adj.* 磁的,有磁性的
stray *adj.* 偶遇的,零散的
negligible *adj.* 可以忽略的

usually uses **a coil of** wire designed specifically to be an inductor. All the **loops'** contribution to the magnetic field adds together to make a stronger field. Unlike capacitors and resistors, practical inductors are easy to make by hand. One can, for instance, **spool** some wire around a short wooden **dowel**, put the spool inside a plastic **aspirin** bottle with the leads hanging out, and fill the bottle with **epoxy** to make the whole thing **rugged**. An inductor like this, in the form of cylindrical coil of wire, is called a **solenoid**. Fig. 1.2 is a **stylized** solenoid. Fig. 1.3 is the symbol used to represent an inductor in a circuit regardless of its actual geometry.

Fig. 1.2 Two common geometries for inductors

Fig. 1.3 The symbol for an inductor

How much energy does an inductor store? The energy density is proportional to the square of the magnetic field strength, which is in turn proportional to the current flowing through the coiled wire, so the energy stored in the inductor must be proportional to I^2. We write $L/2$ for the constant of proportionality, giving

$$E_L = \frac{L}{2} I^2$$

As in the definition of inductance, we have a factor of $1/2$, which is purely a matter of definition. The quantity L is called the inductance of the inductor, and we see that its units must be **joules** per ampere squared. This **clumsy combination** of units is more commonly abbreviated as the **henry**, 1 henry = 1 J/A².

Get to Work

Ⅰ. Matching

Directions: Match the two columns.

1. a capacitor or a condenser   A. an electromotive force
2. microfarad   B. the unit of inductance
3. inductor   C. an electric field to store electrical energy
4. henry   D. a coil of wire with or without a magnetic core
5. emf   E. one millionth of a farad

II. Understanding Checking

Directions: Mark Y (for YES) if the statement agrees with the information given in the text; N (for NO) if the statement contradicts the information given in the text; NG (for NOT GIVEN) if the information is not given in the text.

(　　)1. Instead of batteries, we generally use capacitors and inductors to store energy in oscillating circuits.

(　　)2. A simple condenser consists of two metallic plants separated by a dielectric.

(　　)3. A coil of many turns will have less inductance than one of few turns.

(　　)4. The amount of the stored energy of a charged condenser is proportional to the applied voltage only.

(　　)5. This clumsy combination of units, joules per ampere squared, is more commonly abbreviated as the henry.

III. Passage Completion

Directions: Fill in each of the blanks with one of the words or expressions in the box, making changes if necessary.

| unlike | half | by | storage | two |

When two identical capacitances are placed in parallel, any charge deposited at the terminals of the combined double capacitor will divide itself evenly between the two parts. The electric fields surrounding each capacitor will be __1__ the intensity, and therefore store one quarter of the energy. Two capacitors, each storing one quarter of the energy, give half the total energy storage. Since capacitance is inversely related to energy __2__ , this implies that identical capacitances in parallel give double the capacitance. In general, capacitances in parallel add. This is __3__ the behavior of inductors and resistors, for which series configurations give addition.

This is consistent with the fact that the capacitance of a single parallel-plate capacitor proportional to the area of the plates. If we have __4__ parallel-plate capacitors, and we combine them in parallel and bring them very close together side __5__ side, we have produced a single capacitor with plates of double the area, and it has approximately double the capacitance.

IV. Translation

Directions: Translate English into Chinese.

If two inductors are placed in series, any current that passes through the combined double inductor must pass through both its parts. Thus by the definition of inductance, the inductance is doubled as well. In general, inductances in series add, just like resistances. The same kind of reasoning also shows that the inductance of a solenoid is approximately proportional to its length, assuming the number of turns per unit length is kept constant.

1.3 Voltage Dividers

Learning objectives:

In this section, you will:
- understand the function of a voltage divider.
- comprehend the use of a series dropping resistor R in Fig. 1.4, and the exact resistor value of R.
- explain the disadvantages of the method provided in Fig. 1.4 in actual application.
- find out the values of the resistors in Fig. 1.5.

Warm-Up Activity

Any two-terminal linear network, composed of voltage sources, current sources, and resistors, can be replaced by an equivalent two-terminal network consisting of an independent voltage source in series with a resistor.

Picture 1

(1) Compute the open circuit voltage, U_{oc}.

(2) Compute the Thevenin resistance.

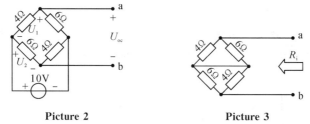

Picture 2　　　　　　　　Picture 3

Text

In many circuits, it is necessary to obtain a voltage not available from the main power source. Rather than have multiple power sources for all needed voltages, we can **derive** other voltages from the main power source. In most cases, the needed voltage is less than the voltage from the main source, so we can use resistors in an **appropriate configuration** to reduce the voltage from the power source, for use in a small circuit.

derive vt. （从……中）得到，获得

appropriate adj. 适当的
configuration n. 构造，结构，配置，外形

If we know precisely both the voltage and current required, we can simply connect a resistor **in series with** the power source, with a value **calculated in accordance with** Ohm's Law. This resistor will drop some of the source voltage, leaving the right amount for the actual load, as shown in Fig. 1.4.

in series with 串联
calculate v. 计算
in accordance with 与……一致，依照

Fig. 1.4 Using a series dropping resistor

Usually, however, this doesn't work too well. The required value of the series **dropping resistor** will almost never be a standard value, and the cost of having special values **manufactured** for specific circuits is **prohibitive**. For example, suppose we have a 9 V battery as main power source, and we want to operate a load that requires 5 V at 3.5 mA. Our series resistor, R, must drop 4 V at 3.5 mA. Using Ohm's Law to calculate the required resistance value, we find that we need a resistance of $4/0.0035\Omega = 1142.8571\ \Omega$ or $1.1428571\ k\Omega$. We have a choice between 1.1 kΩ and 1.2 kΩ as standard 5% values,

dropping resistor 降压电阻器
manufactured adj. 人造的
prohibitive adj. 禁止的，抑制的

but neither will give us what we want.

A more practical solution to the problem is to use two resistors in series, and use the voltage appearing across one of them. This configuration is known as a **voltage divider** because it divides the source voltage into two parts. The basic circuit is shown in Fig. 1.5.

voltage divider
分压器

Fig. 1.5 A voltage divider circuit

In this circuit, the output voltage, U_{OUT}, can be set accurately as a **fraction** of the source voltage, E. Using our example above, we want to select R_1 and R_2 such that we will drop 4 V across R_1, leaving 5 V across R_2. Since U_{OUT} is the voltage across R_2, this will give us the voltage we want.

fraction n. 小部分

We can get a U_{OUT} of 5 V, then, if we set $R_1=1.2$ kΩ and $R_2=1.5$ kΩ. We can also get the same U_{OUT} if we make $R_1=12$ kΩ and $R_2=15$ kΩ. The exact resistor values don't matter, so long as their ratio is correct.

The voltage divider is a very simple circuit that can be highly accurate if not **loaded down**.

load down
使负担过重,超载

Get to Work

Ⅰ. Matching

Directions: Match the two columns.

1. the high-pass filter   A. a device that produces and converts electricity
2. oscillator   B. allows the signals above a certain frequency to pass
3. power supply   C. the ability of a power supply to maintain a constant voltage
4. regulation   D. produce waveforms which may be sinusoidal, square, sawtooth etc.
5. rectifier circuit   E. provide a unidirectional DC

II. Understanding Checking

Directions: Mark Y (for YES) if the statement agrees with the information given in the text; N (for NO) if the statement contradicts the information given in the text; NG (for NOT GIVEN) if the information is not given in the text.

( )1. In most cases, the needed voltage is more than the voltage from the main source.

( )2. If we know precisely both the voltage and current required, we can simply connect a resistor in parallel with the power source.

( )3. The required value of the series dropping resistor will almost never be a standard value.

( )4. The voltage divider is a very simple circuit that can be highly accurate if loaded down.

( )5. Even in those cases where a voltage divider by itself is not sufficient to meet the need, it can serve as the basis of a circuit that will perform as required.

III. Passage Completion

Directions: Fill in each of the blanks with one of the words or expressions in the box, making changes if necessary.

| less | greater | dependent | same | consist |

The voltage divider is the most basic analog input circuit. It __1__ of a variable resistor and a fixed resistor. The output voltage is __2__ on the ratio of the two resistors. For example, when the two resistances are the __3__, the output of voltage is half the input voltage. When the variable resistor is much __4__ than the fixed resistor, the output voltage is closer to the input voltage. When the variable resistor is much __5__ than the fixed resistor, the output voltage is closer to ground.

IV. Translation

Directions: Translate Chinese into English.

1. 晶体管是电子技术中最重要的器件。
2. 有三种基本的逻辑门,它们是与门、或门、非门。
3. 1伏特电压施加在导体上产生了1安培电流,此时电阻为1欧姆。
4. 当把电动势加到线圈上时,线圈中就会产生感应电动势。
5. 从广义上说,把信息从一点传送到另一点就称为通信。

1.4 Filters

Learning objectives:

In this section, you will:
- understand the function of a filter circuit.
- learn about the differences between an active filter and a passive filter.
- master the four groups of filters in terms of their frequency response characteristics.
- analyze how to use the filter in communications.

Warm-Up Activity

Filters can be categorized in terms of their frequency response characteristics.

Picture 1

Picture 2

Picture 3

Picture 4

Text

It is sometimes desirable to have circuits capable of selectively filtering one frequency or range of frequencies out of a mix of different frequencies in a circuit. A circuit designed to perform this frequency selection is called a **filter** circuit, or simply a filter. A filter is a **passive** or **active** circuit designed primarily **for the purpose of** modifying a signal or **source of power**. A passive filter requires no power for its **operation**, and always has a certain amount of loss. An active filter requires its own power, but may provide **gain**.

In power supply, the filter is network of capacitors, resistors and inductors. A good power supply filter will produce a smooth direct current output under variable load conditions, with good **regulation**.

Filters are used in **communications** practice to **eliminate** energy at some frequencies while allowing energy at other frequencies to pass with little or no **attenuation**. Such filters, if passive, are constructed using capacitors, inductors and sometimes resistors. Active filters generally utilize **operational amplifiers**.

Filters can be **categorized**, **in terms of** their frequency response characteristics, into four groups. The **band-pass filter** allows signal between two predetermined frequencies to pass, but attenuates all other frequencies. The **band-rejection or band-stop filter** eliminates all energy between two frequencies, and allows signals outside the limit frequencies to pass. The **high-pass filter** only allows the signals above a certain frequency to pass. The **low-pass filter** only allows the signal below a certain frequency to pass. The actual response characteristics of all four types of filter are highly variable, and depend on many different factors.

---

filter *n.* 滤波器
passive *adj.* 无源的
active *adj.* 有源的
for the purpose of 为了
source of power 电源
operation *n.* 运转, 操作, 工作
gain *n.* 增益
regulation *n.* 稳压
communication [计] 通信
eliminate *vt.* 排除, 消除
attenuation *n.* 衰减
operational amplifier 运算(操作)放大器
categorize *v.* 分类
in terms of 根据, 按照
band-pass filter 带通滤波器
band-rejection or band-stop filter 带阻滤波器
high-pass filter 高通滤波器
low-pass filter 低通滤波器

Get to Work

Ⅰ. Matching

Directions: Match the two columns.

1. the high-pass filter

2. a filter

3. the low-pass filter

4. the band-pass filter

5. the band-rejection filter

A. allows the signals between two predetermined frequencies to pass

B. allows the signals above a certain frequency to pass

C. eliminates all energy between two frequencies

D. allows the signals below a certain frequency to pass

E. a passive or active circuit designed primarily for the purpose of modifying a signal or source of power

Ⅱ. Understanding Checking

Directions: Mark Y (for YES) if the statement agrees with the information given in the text; N (for NO) if the statement contradicts the information given in the text; NG (for NOT GIVEN) if the information is not given in the text.

(　　)1. A passive filter requires power for its operation, and always has a certain amount of loss.

(　　)2. In power supply, the filter is network of capacitors, resistors and inductors.

(　　)3. Filters can be categorized, in terms of their frequency response characteristics, into four groups.

(　　)4. The band-rejection filter allows signal between two predetermined frequencies to pass, but attenuates all other frequencies.

(　　)5. The actual response characteristics of all four types of filter are highly variable, and depend on many different factors.

Ⅲ. Passage Completion

Directions: Fill in the blanks with words according to the text.

1. A filter is a ＿＿＿＿＿ or ＿＿＿＿＿ circuit designed primarily for the purpose of modifying a signal or source of power.

2. In power supply, the filter is network of ＿＿＿＿＿, ＿＿＿＿＿ and ＿＿＿＿＿.

3. Filters can be categorized, in terms of their ＿＿＿＿＿ characteristics, into four groups.

4. The high-pass filter allows only the signals ＿＿＿＿＿ a certain frequency to pass.

Ⅳ. Translation

Directions: Translate Chinese into English.

1. 无源滤波器由电阻、电容、电感组成。

2. 有源滤波器由集成运放组成。
3. 滤波器按其频率特性可分为高通滤波器、低通滤波器、带通滤波器和带阻滤波器。
4. 滤波器主要用于滤除某些频率成分而允许另一些频率成分无衰减地通过。
5. 滤波器包括有源滤波器和无源滤波器。

CAREER SPEAKING

Job Interview

Ⅰ. Warm-Up

Model

(A=Applicant　　I=Interviewer)

I: Mr. Zhang, I have read your resume. Why did you choose to major in mechanical engineering?

A: Many factors led me to major in mechanical engineering. The most important factor is I like *tinkering with* machines.

I: What are you primarily interested in about mechanical engineering?

A: I like designing products, and one of my designs received an award. Moreover, I am familiar with CAD.

I: Why did you decide to apply for this position?

A: Your company has a very good reputation, and I am very interested in the field your company is in.

I: What do you think determines an employee's progress in a company such as ours?

A: Interpersonal and technical skills.

I: We have several applicants for this position. Why do you think you are the person we should choose?

A: I have the abilities, qualities and experience that you requested in your job advert. For example, I have three years' experience in designing products and I got leadership experience while serving the college student union as the president.

I: That sounds very good. How do you see your career development?

A: After a few years of gaining experience in the company and furthering my professional qualifications I'd like to put my experience and skills to use in management. I want to become a supervisor in your R&D (research and design) department.

Note

tinker with   摆弄,修理

II. Matching

Directions: The following conversation is based on a job interview. "I" stands for the interviewer while "A" represents the applicant. Please match each question with its answer, then play roles with your partner.

I: Have you obtained any certificates of technical qualifications or licenses?
A: _____.
I: How many years have you had the driving license?
A: _____.
I: That's good. What special skills do you have?
A: _____.
I: What computer languages have you learned?
A: _____.
I: Have you gotten any special training in programming?
A: _____.
I: How do you think of your proficiency in written and spoken English?
A: _____.

> S1: I have two years' driving experience.
> S2: I have experience in computer operation, proficiency in Microsoft Windows, Microsoft Word and Microsoft Excel.
> S3: I have learned English for 10 years, and I have passed College English Test Band 4 and Band 6. My spoken English is fairly good enough to express myself fluently.
> S4: No, but I have database programming experience and network knowledge.
> S5: Yes, I have received an Engineer Qualification Certificate and a driving license.
> S6: Visual C++ and C.

III. Closed Conversation

Directions: The following conversation is based on a job interview. "I" stands for the interviewer while "A" represents the applicant. Please fill the information in the blanks according to the Chinese, then play roles with your partner.

I: What kind of person do you think you are?
A: _____ (我觉得自己精力充沛,做事有热情).
I: What are your weaknesses?
A: _____
(有时为了把事情办得完美些,我让自己背上太多的压力).
I: How would your friends or colleagues describe you?
A: _____
(他们认为我是很友好、敏感、关心他人和有决心的人).
I: What personality do you admire?
A: _____ (我欣赏诚实、灵活而且容易相处的人).
I: How do you spend your leisure time?
A: _____ (我爱玩游戏和体育运动).
I: What qualities would you expect of persons working as a team?
A: To work in a team, in my opinion, _____
(合作精神和进取精神两者皆不可少).

Ⅳ. Semi-Open Conversation

Directions: The following conversation is based on a job interview. "I" stands for the interviewer while "A" represents the applicant. Please fill in the blanks according to your own information, then play roles with your partner.

I: Which school are you attending?
A: _____.
I: When will you graduate from that university?
A: _____.
I: What is your major?
A: _____.
I: How did you get on with your studies at university?
A: _____.
I: Which course did you like best?
A: _____.
I: Did you get any honors and awards at college?
A: _____.

Ⅴ. Live Show

Challenge 1

Directions: You, a new graduate from a vocational college, are hunting a job. Please act it out with your partner or partners. The model and the useful expressions below are just for your reference.

Challenge 2

Directions: You, an employee in an industry laser company, are not satisfied with your current salary and benefits and decide to talk to your boss. Please act it out with your partner or partners. The model and the useful expressions below are just for your reference.

Model

(A=Applicant   I=Interviewer)

A: Good morning. I came in answer to your advertisement for a salesperson.

I_1: I see. What experience do you have?

A: I'm afraid I haven't had any experience in just this sort of work.

I_2: Have you got any selling experience at all?

A: I worked in a department store in a small town last summer vacation.

I_1: Now tell us about your educational background.

A: I graduated from Hezhe Vocational College. I majored in Optoelectronic Technology.

I_2: What courses have you completed?

A: The courses I completed are Circuits, C Language, Program Principles and Application of Microcontrollers, Laser Principles, Photoelectric Detection Technology, Optical Fiber Communication and so on.

I_1: Why would you like to work with us?

A: It's an interesting job, and your company is one of the best-known.

I_2: Do you know anything about this company?

A: Yes, a little. Yours is an American company. As far as I know, your company is a world-famous one which produces cosmetics and skincare products.

Useful expressions

Why do you think you are qualified for this position?

I have worked at CBA Company for 4 years and got a lot of practical experience.

What's your technical title now?

I am a senior mechanical design engineer.

What made you decide to change your job?

Because I want to change my working environment and seek new challenges.

What do you think is the most important qualification for a salesperson?

I think it is confidence in oneself and the products.

Do you have any questions to ask about this job?

I'd like to know something about the salary.

WRITING

Resume

求职简历是求职者生活、学习经历及职位期望等的简述,一般以大纲或一览表的形式呈现,其目的是让招聘单位尽快地了解求职者的基本情况。英文简历通常包括下列内容。

● Job Objective(求职目标)
● Personal Data(个人资料)
Name(姓名)　　　　　Sex(性别)　　　　　Date of Birth(出生日期)
Place of Birth(出生地点)　Native Place(籍贯)　Marital Status(婚姻状况)
Nationality(民族/国籍)　Height(身高)　　　Weight(体重)
Health(健康状况)　　Present Address(目前地址)　Permanent Address(永久地址)
Telephone No.(电话号码) E-mail(电子邮箱)
● Education(所受教育)
● Work Experience(工作经历)
● Awards & Organizations(所受奖励和社会活动)
● Hobbies/Specialties(个人爱好或特长)
● References(证明人)
以上部分项目可视情况删去。

Sample 1

PERSONAL DATA

Name: Edward Brown　　　　Sex: Male　　　　Date of Birth: Aug. 20, 1985
Height: 178cm　　　　　　　Weight: 70kg　　　Health: Excellent
Address: 32 Gilmore Lane, Boston, Massachusetts 02101
Telephone No.　(617) 555-1212　　　　　　　E-mail: Whiteman@yahoo.com

JOB OBJECTIVE

Director of Banking Training

EDUCATION

West Springfield University, Illinois
Master of Business Administration (MBA), June 2009

West Springfield University, Illinois
Bachelor of Finance, June 2006

WORK EXPERIENCE

August 2008—Present Standard Chartered Bank, Springfield, Illinois
Data Services Manager/Administrative Assistant
Supervise, manage and train 22-person office staff
Prepare for meetings and correspond with member representatives on upcoming meetings

June 2006—July 2008 Standard Chartered Bank, Springfield, Illinois
Bookkeeper
Received cash and check receipts, maintained the ledger book and computer record of bank deposits
Wrote and distributed employee and contractor checks

SKILLS

Microsoft Word 2007
Microsoft Excel 2007

HONORS

Model Worker in Standard Chartered Bank (Springfield Branch) in 2010
West Springfield University Trustee Scholarship Recipient in 2008

HOBBIES

Cycling, Traveling, Fishing

References Provided upon request

Sample 2

Full Name: Li Jing
Address: No. 113 Wuda Road, Ximing City
Mobile: 16700781148
Date of Birth: March 5, 1990
Sex: Female
Nationality: Chinese
Marital Status: Single
Health: Excellent
Job Objective: A technician in a laser company with opportunity for advancement

Education: 2008—2011　Changshan Vocational College
　　　　　　　　Major: laser processing
　　　　2005—2008　No. 2 High School, Hushui County
Part-Time Work Experience: July 2009—August 2009　A salesperson in Jingming Computer Company
　　　　　　　　　　　　　　July 2010—August 2010　A worker operating laser cutting in Feida Laser Co. , Ltd.
Awards & Honors: The National Scholarship Recipient in 2010
　　　　　　　　　　"Excellent Student" of Changshan Vocational College in 2009
Strengths: being good at communication and cooperation
　　　　　　　excellent team spirit and professional ethics
Interests: watching movies, filming and listening to music
References: available upon request

Challenge

Directions: Finish a resume for yourself in terms of your own conditions and situations.

EXTENSIVE READING

1.5　Ohm's Law

One thing we need to be able to do when we see a schematic circuit diagram is to perform mathematical calculations to define the precise behavior of the circuit. All information required to perform such calculations should be included on the schematic diagram itself. That way the information is all in one place, and any required detail can be determined readily.

Consider the basic circuit shown to the right. We know immediately that the battery voltage is 6 V and that the resistor is rated at 1000 Ω. Now, how can we determine how much current is flowing through this circuit?

If you go back to basic circuit concepts, you'll note that the relationship between voltage, current, and resistance is given as $E=IR$. Using basic algebra we can also rewrite this as:

$$R=E/I$$
$$I=E/R$$

These three equations describe Ohm's Law, which defines this relationship.

In the circuit shown above, we see that $E=6$ V and $R=1000$ Ω. To find the current flowing in this circuit, we must select the equation that solves for *I*. Using that equation, we note that:

$$I = E/R$$
$$I = 6V/1000\Omega$$
$$I = 0.006 \text{ A} = 6 \text{ mA}$$

All calculations involving Ohm's Law are handled in exactly the same way. If the circuit gets complex, the calculations must be tailored to match. However, each calculation is still just this simple.

Exercise

Directions: Choose the best answers from the four choices.

1. The flow of electrons through a conductor is called _____.
   A. voltage   B. current
   C. resistance   D. Ohm's Law
2. The relationship between volts, amperes and ohms can be represented by _____.
   A. voltage   B. current
   C. resistance   D. Ohm's Law

1.6 What Is Electricity

The modern science of electricity originated with Benjamin Franklin, who began studying and experimenting with it in 1747. In the course of his experiments Franklin determined that electricity was a single force, with positive and negative aspects. Up to that point, the prevailing theory was that there were two kinds of electricity: one positive, the other negative.

To describe his experiments and results, Franklin also coined some twenty-five new terms, including armature, battery, and conductor. His famous kite-flying experiment in a thunderstorm was performed in 1752, near the end of his work in this field.

Since then, many scientists, inventors, and entrepreneurs around the world have performed their own experiments, verifying and building on Franklin's beginnings in the field. Now, some 250 years later, we use electricity in almost every aspect of our daily lives. In some cases, we may not even realize that electricity is involved as an integral part of our activities.

So just what is electricity? Let's start with the dictionary definition, to give all of us some common ground. The American Heritage Dictionary actually gives four specific definitions.

Electricity

1. The class of physical phenomena arising from the existence and interactions of electric charge.
2. The physical science of such phenomena.

3. Electric current used or regarded as a source of power.
4. Intense emotional excitement.

We will skip the fourth definition as having no useful connection to the other three, and deal with electricity as a physical phenomenon which may be studied and manipulated using the tools of science.

When Franklin developed his hypotheses about electricity, he arbitrarily assumed that the actual carriers of electrical current had a positive electrical charge. All of his theories, calculations, and descriptions were based on this assumption. Fortunately, his experiments still worked even with this incorrect assumption built into them. This "conventional" assumption was used for 200 years or more, and is still built into many of the common rules and procedures used to design and analyze electrical devices and behaviors.

We now know that the actual carriers of electricity are electrons, which have a negative electrical charge as defined in our system of science. Because of this, "electron theory" has been replacing "conventional theory" in schools and in regular usage.

The obvious next questions are:
1. What is an electron?
2. Where do electrons come from?
3. How do they carry an electrical charge from place to place?

We'll begin answering those questions when we take a closer look at electrons.

Exercise

Can you tell us what electricity is in your opinion?

TRANSLATION

Features of Scientific English

科技英语的特点

科技英语不同于日常英语、文学英语等,具有自身独有的规律和特点,主要体现在以下几个方面。

1. 被动语句多

科技英语的主要作用是阐述科学事实,描述发展变化,揭示本质特征,论述中应保持客观中立的态度,强调准确性和可信度。因此尽量使用第三人称叙述,采用被动语态,避免主观色彩。例如:

The potential of fiber optics in other areas ***has been realized***.

人们已经开始认识到光纤在其他领域的应用潜力。

(has been realized 是被动语态,翻译为"人们已经开始认识到",转化为主动语态。)

2. 名词化结构多

名词化结构可以代替从句,使语言更简练,结构更紧凑。例如:

A *sound understanding of* machine functions and performance is the basis for safe, accident-free and injury-free operation.

彻底了解机器的功能和性能是安全操作、避免发生事故，以及防止受伤的前提。

（句中 a sound understanding of 是名词化结构。）

3. 非谓语动词多

大量使用分词、动名词、不定式、独立主格结构代替从句，也是为了缩短句子，使表达更加醒目直观、简明清晰。例如：

A sinusoidal oscillator generates a signal ***having*** a sine waveform.

正弦振荡器发出正弦波形的信号。

（having 是现在分词，作后置定语，代替定语从句。）

The article deals with microwaves, ***with particular attention paid*** to radio location.

这篇文章是研究微波的，其中特别注意无线电定位问题。

（with particular attention paid 是分词复合结构，在句中做补充说明。）

The power station is automatic, ***the sluices being worked*** form the control board.

这个发电站是自动化的，闸门可以由控制板操纵。

（the sluices being worked 是现在分词的被动形式作独立主格结构，在句中作状语。）

4. 词类转换多

英语单词有不少词是一词多性，既可以作名词，又可以作动词、形容词、介词或副词。还有很多词一词多义，而且在不同的场合、不同的语境、不同的专业中意思也不相同。例如：

clear

adj. 明朗的 a clear day（晴天）

　　　清楚的 a clear photograph（清晰的照片）

　　　无罪的 a clear conscience（无愧的良心）

　　　完全的 a clear victory（彻底的胜利）

　　　无瑕疵的 a clear complexion（无瑕的肤色）

　　　无障碍的 a clear view（一览无余的视野）

adv. 清楚地 speak clear（讲话清楚）

　　　完全 He got clear away.（他逃得无影无踪。）

n. ［建］中空体内部的尺寸 The tube is ten feet wide in the clear.（这管子内径宽十英尺。）

vt. 清除 clear the plates away（把盘子拿走）

　　　跳过 The horse cleared the fence.（马跳过了栅栏。）

　　　净得 He cleared two thousand dollars last year.（去年他净赚了 2 000 美元。）

　　　为船只结关 clear a ship at the customhouse（在海关办理船只的出港手续）

vi. 变明朗 The weather has cleared up.（天放晴了。）

5. 短语结构多

例如：

Most of the products ***on display*** are new ones.

展出的产品大多数都是新产品。

（短语 on display 作定语。）

This machine is **out of order**.

这台机器失灵了。

（短语 out of order 作表语。）

6. 复杂长句多

科技文章要求叙述准确，用词严谨，因此一句话里常常包含多个分句，这种复杂且长的句子居科技英语难点之首，阅读、翻译时要按汉语习惯来加以分析，以短代长，化难为易。例如：

An uncharged object contains a large number of atoms, each of which normally contains an equal number of electrons and protons, but with some electrons temporarily free from atoms.

不带电的物体含有大量原子，每个原子通常含有等量的电子和质子，但是有一些电子暂时脱离了原子的束缚。

（这是一个带有非限定性定语从句和"with＋名词＋形容词"的复合结构组成的长句，定语从句和复合结构对句子起补充说明作用。）

TASK

Basic Circuit Concepts

Target	Understand and describe the basic circuit.
Requirements	Make an English presentation according to lesson 1.1. Produce a micro-video to express professional knowledge. Add accurate English subtitles to the video.
Contents	Basic circuit concepts: Direct current (DC) Voltage current resistance The relationship between voltage, current, and resistance in an electrical circuit
Group:_____	Project	Name	Software	Time	Score
	Material collection				
	Script				
	PPT				
	Reader				
	Subtitle				
	MP4				
Operating environment	XP/Win7/Win8/ Mobile phone
Product features	

Script reference

标题	专业内容(讲稿与英文字幕)	PPT中需显示的文字	表现形式	素 材
每个知识点的要点或者小节名字	做报告时的英文讲稿 视频中的英文字幕	需要特别提醒的文字	讲解到某个地方或某句话时,出现的图片、动画或者文字	图片或者视频
Contents	**Basic circuit concepts** Contents: Direct current (DC) Voltage current resistance The relationship between voltage, current, and resistance in an electrical circuit	Basic circuit concepts: Direct current (DC) Voltage current resistance The relationship between voltage, current, and resistance in an electrical circuit	文字	
1. A basic DC circuit	Figure 1.1 shows the basic type of an electrical circuit, in the form of a block diagram. It consists of a source of electrical energy, some sort of load to make use of that energy, and electrical conductors connecting the source and the load.	Source Load Conductors	图文	图1 基本电路框图
	The electrical source has two terminals, designated positive (+) and negative (−). As long as there is an unbroken connection from source to load and back again as shown here, electrons will be pushed from the negative terminal of the source, through the load, and then back to the positive terminal of the source. The arrows show the direction of electron current flow through this circuit. Because the electrons are always moving in the same direction through the circuit, their motion is known as a direct current (DC).	Two terminals An unbroken connection The arrows	图示	图1 基本电路框图

Unit 1 Basic Circuit and Its Applications   29

续表

标题	专业内容（讲稿与英文字幕）	PPT 中需显示的文字	表现形式	素 材
Source	The source can be any source of electrical energy. In practice, there are three general possibilities: it can be a battery, an electrical generator, or some sort of electronic power supply.		图示	图 2 电池 图 3 发电机
Load	The load is any device or circuit powered by electricity. It can be as simple as a light bulb or as complex as a modern high-speed computer.		图示	图 4 灯泡 图 5 电脑
2. Definitions	The electricity provided by the source has two basic characteristics, called voltage and current. These are defined as follows:		文字	
	Voltage The electrical "pressure" that causes free electrons to travel through an electrical circuit. Also known as electromotive force (emf). It is measured in volts.	Voltage	定义	图 6 公式 $E = I \times R$
	Current The amount of electrical charge (the number of free electrons) moving past a given point in an electrical circuit per unit of time. Current is measured in amperes.	Current	定义	图 6 公式
	Resistance That characteristic of a medium which opposes the flow of electrical current through itself. Resistance is measured in ohms.	Resistance	定义	图 6 公式

续表

标题	专业内容（讲稿与英文字幕）	PPT中需显示的文字	表现形式	素材
3. Relationships	The relationship between voltage, current, and resistance in an electrical circuit is fundamental to the operation of any circuit or device.	The relationship between voltage, current, and resistance in an electrical circuit is fundamental to the operation of any circuit or device.	文字	
	Verbally, the amount of current flowing through a circuit is directly proportional to the applied voltage and inversely proportional to the circuit resistance.	Verbally		
	By explicit definition, one volt of electrical pressure can push one ampere of current through one ohm of resistance. Two volts can either push one ampere through a resistance of two ohms, or can push two amperes through one ohm.	Explicit definition		
	Mathematically, $E = I \times R$, where E = The applied voltage, or EMF I = The circuit current R = The resistance in the circuit	Mathematically	图文	图6公式

UNIQUE ENGLISH

Brain Teaser

1. What makes a road broad?
2. Who makes a million dollars a day?
3. What is the end of everything?
4. What can pierce one's ears without a hole?
5. Why do people go to bed?
6. Why wasn't he hurt after falling off a 1,000-step stair?

Unit 2

Optics

In this unit, you are going to learn the following contents:

READING IN DEPTH

2.1　Nature of Light

2.2　Reflection and Refraction

2.3　Numerical Aperture (NA)

2.4　Electromagnetic Radiation

CAREER SPEAKING

Meeting Guests

WRITING

Letter of Thanks

EXTENSIVE READING

2.5　What Is Light

2.6　Convex Lens

TRANSLATION

Translation of Words and Phrases

TASK

Electromagnetic Radiation

UNIQUE ENGLISH

Incredible Sentences

READING IN DEPTH

2.1 Nature of Light

Learning objectives:

In this section, you will:
- understand the range of visible wavelengths.
- learn about the relationship between light and radio waves.
- comprehend the electron volts.
- learn about wave nature of light.
- explain the particle theory.

Warm-Up Activity

Do you know the names of the apparatus below?

Picture 1

Picture 2

Picture 3

Picture 4

Text

Although light **pervades** human existence, its fundamental nature remains at least a partial mystery. We know how to **quantify** light phenomena and make predictions, based on this knowledge, and we know how to use and control light for our own convenience. Yet light is often interpreted in different ways to explain different experiments and observations: sometimes light behaves as a **wave**, sometimes light behaves as a **particle**.

Wave Nature of Light

Many light phenomena can be explained if we look at light as being an electromagnetic wave having a very high **oscillation** frequency and a very short wavelength. We use the **term** optic (as well as the term light) to refer to frequencies in the infrared, visible, and ultraviolet portions of the spectrum. We do this because so many of the same analyses, techniques, and devices are applicable to these ranges.

The range of frequencies (or wavelengths) that primarily interests us is introduced here. Visible wavelengths extend from 0.4 μm (which we distinguish as the color blue) to 0.7 μm (which appears to us as red). **Silica** glass fibers are not very good transmitters of light in the visible region. They attenuate the waves to such an extent that only short transmission links are practical. Losses in the ultraviolet are even greater. In the infrared, however, these regions in which glass fibers are relatively efficient transmitters of light. These regions occur at wavelengths around 0.85 μm and between 1.26 μm and 1.75 μm. These wavelengths are sometimes referred to as the fiber transmission windows.

Although light waves have much higher frequencies than **radio waves**, they both obey the same laws and share many characteristics. All electromagnetic waves have electric and magnetic fields associated with them, and they all travel very quickly.

In empty space (usually referred to as free space), electromagnetic waves travel at the **velocity** 3×10^8 m/s. This velocity, indicated by c, is appropriate for wave travel in the atmosphere. In solid media,

pervade v. 遍及
quantify v. 量化

wave n. 波
particle n. 粒子

oscillation n. 摆动,振动
term n. 术语

silica n. [化]硅石

radio wave 无线电波

velocity n. 速度,速率

the wave velocity differs. Its value depends on the material and on the geometry of any wave guiding structure that is present. The wavelength of a light beam is given by

$$\lambda = v/f \tag{2.1}$$

where v is the **beam** velocity and f is its frequency. The frequency is determined by the emitting source and does not change when the light travels from one material to another. Instead, the velocity difference causes a change in wavelength according to Eq. (2.1).

As an example, consider radiation at 0.8 μm. Using Eq. (2.1), with $v=c$, yields the frequency 3.75×10^{14} Hz. This is a fast oscillation indeed. The period of this oscillation (the reciprocal of its frequency) is then 2.67×10^{-15} s, an extremely short time span. We should also note that the wavelengths of optic beams are of the order of 1 μm near the visible spectrum. Optic wavelengths are so small that most devices used in a fiber system have dimensions of many wavelengths. This is unlike the situation at lower frequencies, where device sizes can be a wavelength or less.

Particle Nature of Light

So far, we have looked at light as being a wave. Sometimes light behaves unlike a wave and instead behaves as though it were made up of very small particles called **photons**. The energy of a single photon is

$$w_p = hf \tag{2.2}$$

where $h = 6.626 \times 10^{-34}$ J·s and is called Planck's constant. The energy computed from Eq. (2.2) has the units joules. It is impossible to break light into divisions smaller than the photon. Ordinarily, beams of light contain huge numbers of photons. The following example illustrates this.

beam n. 光线，（粒子的）束

photon n. [物]光子

Example

Find the number of photons incident on a detector in 1 s if the optic power is 1 μW and the wavelength is 0.8 μm.

Solution

From Eqs. (2.1) and (2.2), the energy of a single 0.8 μm photon is

$$W_p = hf = hc/\lambda = 2.48 \times 10^{-19} \text{ J}$$

Because power is the rate at which energy is delivered, we can write the total energy

$$W = Pt$$

Multiplying the power, 1 μW, times the 1 s time interval yields the energy 1 μJ.

The number of photons required to make up 1 μJ is
$$\frac{W}{W_p} = \frac{10^{-6}}{2.48 \times 10^{-19}}$$
$$= 4.03 \times 10^{13}$$

In the example above, if we reduced our observation time to as little as 1 ns, we would still receive over 4000 photons. The most sensitive receivers can detect the presence of radiation when only a few photons arrive.

A convenient unit of energy for photons is the electron volt (eV). It is the kinetic energy, acquired by an electron when it is accelerated by 1 V of potential difference. The relationship between electron volts and joules is
$$1 \text{ eV} = 1.6 \times 10^{-19} \text{ J}$$

The 0.8 μm photon treated in example has an energy in electron volts given by
$$2.48 \times 10^{-19} \text{ J} / 1.6 \times 10^{-19} \text{ J/eV} = 1.55 \text{ eV}$$

Particle theory explains generation of light by sources, such as light-emitting diodes, lasers, and laser diodes. It also explains detection of light by conversion of optic radiation to electrical current.

Get to Work

Ⅰ. Matching

Directions: Match the words or expressions in the left column with the Chinese equivalents in the right column.

1. radio wave   A. 电磁波
2. oscillation frequency   B. 振荡频率
3. photon   C. 焦耳
4. electromagnetic wave   D. 光子
5. joule   E. 无线电波

Ⅱ. Understanding Checking

Directions: Mark Y (for YES) if the statement agrees with the information given in the text; N (for NO) if the statement contradicts the information given in the text; NG (for NOT GIVEN) if the information is not given in the text.

(   )1. Sometimes light behaves as a wave, sometimes light behaves as a particle.
(   )2. Many light phenomena can be explained if we look at light as being an electromagnetic wave having a very high oscillation frequency and a very short wavelength.
(   )3. Electromagnetic waves travel at the velocity 3×10^3 m/s all the time.
(   )4. A convenient unit of energy for photons is the electron volt (eV).

(   )5. The relationship between electron volts and joules is 1 eV=1.6×10^{-19}J.

III. Passage Completion

Directions: Fill in each of the blanks with one of the words or expressions in the box, making changes if necessary.

| of | with | assumption | nature | that |

For a long time, physicists tried to explain away the problems with the classical theory of light as arising from an imperfect understanding of atoms and the interaction of light with individual atoms and molecules. The ozone paradox, for example, could have been attributed to the incorrect __1__ that the ozone layer was a smooth, continuous substance, when in reality it was made __2__ individual ozone molecules. It wasn't until 1905 that Albert Einstein threw down the gauntlet, proposing __3__ the problem had nothing to do __4__ the details of light's interaction with atoms and everything to do with the fundamental of __5__ light itself.

IV. Translation

Directions: Translate the following sentences from Chinese into English.
1. 可见光的波长范围是从0.4 μm 到 0.7 μm。
2. 光波同时具有波和粒子的双重性质。
3. 真空中电磁波的速度是3×10^8 m/s。
4. 通常光束是由大量光子组成的。
5. 电子伏特是一个便于计量光子能量的单位。

2.2 Reflection and Refraction

Learning objectives:
In this section, you will:
● understand the reflection phenomenon.
● learn about the definition of incident light and reflected light.
● learn about the angle of incidence and the angle of reflection.
● master the relationship between the angle of incidence and the angle of reflection.
● explain the refraction phenomenon when light travels from air into water.

Warm-Up Activity

Please enjoy the scenery.

Picture 1

Picture 2

Picture 3

Picture 4

Picture 5

Text

In talking about the fundamental nature of light, we **indicated** that light tends to travel in a straight line, unless it is acted on by some external force or condition. The obvious next question is, "What kinds of forces or conditions can affect light, and how?"

To answer this question, we start with what we can see in our daily life. For example, we already know that light won't pass through the wall of a house, but it will go through a window. **Furthermore**, some windows **introduce** noticeable distortion in what we see. Looking carefully at the glass of such a window, we can see that

indicate *vt.* 指出，显示，象征，预示

furthermore *adv.* 此外，而且
introduce *vt.* 传入，引进

it is uneven, perhaps with **ripples** across its surface.

We have also seen that some surfaces show accurate **images** of an actual **scene** nearby, while other surfaces show distorted images, but most surfaces only show one or more colors no matter how we look at them.

On top of that, some surfaces seem very much darker than their surroundings, while other surfaces seem just as bright as their surroundings. Indeed, occasionally you can see something that looks brighter than its surroundings, both at night and in broad daylight.

A. Reflection

When light reflects off a surface, it follows some rather basic rules which have been gradually determined by observation. A ray of light **approaches** a reflecting **horizontal** surface at an angle of 45°, bounces from the surface, and leaves at an angle of 45°.

So that we can agree fully on what we are talking about, we need to define a few terms below.

Incident Light

Light approaching a surface is known as incident light. This is the incoming light before it has reached the surface.

Reflected Light

After light has struck a surface and **bounced** off, it is known as reflected light. This is the light that is now departing from the surface.

Angle of Incidence

The angle at which a ray of light approaches a surface, reflective or not, is called **the angle of incidence**. It is measured from an imaginary line perpendicular to the plane of the surface in question to the incoming ray of light.

Angle of Reflection

Once the light has reflected from a reflective surface, the angle at which the light departs from the surface is called **the angle of reflection**. This angle is also measured from a **perpendicular** to the reflecting surface to the departing ray of light.

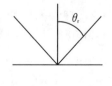

When light reflects from a surface, the angle of reflection is always equal to the angle of incidence.

B. Refraction

Refraction is the name given to the observed phenomenon that

---

ripple *n.* 波纹
image *n.* 图像
scene *n.* 现场，场面，情景，景色

approach *v.* 靠近
horizontal *adj.* 水平的

bounce *v.* （使）反跳，弹起

the angle of incidence
入射角

the angle of reflection
反射角
perpendicular *n.* 垂线
adj. 垂直的，正交的

light changes direction, or "bends", as it passes the **boundary** between one **medium** and another. This is shown to the below, in a general sense.

boundary *n.* 边界, 分界线

medium *n.* 媒介

Here, we see a beam of light traveling through air, until it meets a pool of water. It arrives at some angle to the surface as shown. As it passes through the boundary, going from air into water, it actually slows down. Since even a single ray of light has a **finite** thickness, the part that enters the water first slows down first, causing the light ray to change direction to a steeper angle in the water.

finite *adj.* 有限的, [数]有穷的, 限定的

If we change the angle at which the light enters the water, we find that the angle of the light in the water also changes, such that we see no change at all if the light source is directly overhead so that the entering ray of light is perpendicular to (in mathematical terms be normal to) the surface. As we change the entering angle more and more away from the perpendicular, we see that the ray of light in the water has bent more and more away from the direction taken by that ray of light in the air.

Get to Work

I. Matching

Directions: Match the words or expressions in the left column with the Chinese equivalents in the right column.

1. incident light   A. 折射
2. reflected light   B. 入射光
3. refraction   C. 水平的
4. horizontal   D. 反射光
5. vertical   E. 垂直的

Ⅱ. Understanding Checking

Directions: Mark Y (for YES) if the statement agrees with the information given in the text; N (for NO) if the statement contradicts the information given in the text; NG (for NOT GIVEN) if the information is not given in the text.

(   )1. Light always travels in a straight line.

(   )2. Incident light is the incoming light before it has reached the surface.

(   )3. Light approaching a surface is known as reflected light.

(   )4. Both angle of reflection and angle of incidence are measured from a line perpendicular to the surface.

(   )5. Light can pass through the wall of a house.

Ⅲ. Passage Completion

Directions: Fill in each of the blanks with one of the words or expressions in the box, making changes if necessary.

refraction	process	although	various	different
with	psychological	definition	category	for example

Physical optics is concerned with the creation, nature, and properties of light. __1__ optics pertains to the role of light in vision. Geometrical optics deals __2__ the properties of reflection and __3__ of light, as part of the study of mirrors, lenses, and optical fibers.

As you can see from the dictionary __4__ above, the study of optics falls into three general __5__: light itself, what it is and how it behaves; how we perceive light through the sense of sight; and how light can be manipulated through such __6__ as reflection and refraction. __7__ we talk of these categories as if they were separate and distinct, actually the whole reality of sight and vision constantly involves all of these categories in __8__ combinations.

__9__, consider a rainbow. This visual phenomenon not only involves both reflection and refraction, but also is a practical demonstration that different colors of light refract, __10__ even under fixed circumstances.

Ⅳ. Translation

Directions: Translate sentences from Chinese into English.

1. 当光线从表面反射时,它遵循一些观察得到的基本规则。

2. 到达一个表面然后反弹的光,称为反射光。

3. 光从一种物质到另一种物质,在界面处光的方向发生改变或"弯曲"的现象称为折射。

4. 通常光束是由大量光子组成的。

5. 当入射光越来越远离铅垂线时,我们可以看到光在水中的方向越来越偏离入射光在空气中的方向。

2.3 Numerical Aperture (NA)

Learning objectives:
In this section, you will:
● understand the meaning of numerical aperture.
● learn about the major advantage of the graded-index multimode fibers.
● comprehend the calculating equation of NA.

Warm-Up Activity

Please describe the condition of total reflection.

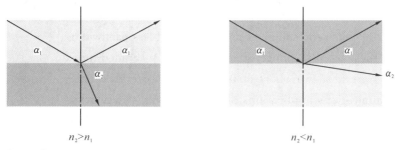

Phrases for reference: total reflection; a vacuum, index of refraction ($n=1$); the up media with index of refraction n_1, the media below with index of refraction n_2

Text

In optics, the numerical aperture (NA) of an optical system is a dimensionless number that characterizes the range of angles over which the system can accept or emit light. By incorporating index of refraction in its definition, NA has the property that it is constant for a beam as it goes from one material to another, provided there is no optical power at the interface. The exact definition of the term varies slightly between different areas of optics. NA is commonly used in microscopy to describe the acceptance cone of an objective (and hence its light-gathering ability and resolution), and in fiber optics, in which it describes the range of angles within which light that is incident on the fiber will be transmitted along it.

NA is a **critical performance specification** for multimode fibers. It indicates the maximum angle at which a **particular** fiber can accept the light that will be transmitted through it. The higher an optical fiber's NA, the larger the cone of light that can be coupled into its core.

Graded-index multimode fibers have a large NA. This is a major advantage of the product: it enables them to be used with relatively low-cost optical components and light sources such as light-emitting diodes (LEDs) and **Vertical Cavity Surface Emitting Lasers** (VCSELs). LEDs and VCSELs, which have large spot sizes, can be easily coupled to multimode fibers. In contrast, single-mode fibers, which have a small NA, typically use narrow width lasers as power sources and carry only one mode of light straight through a very narrow core. Transmitter **alignment** and **tolerances** must be very precise to couple the small beam of light into the tiny core of a single-mode fiber. This drives up the cost of single-mode components.

Multimode fibers allow more modes of light to be transmitted, resulting in greater pulse spreading, or dispersion, and less bandwidth. Consequently, these easily-connectorized, high-NA graded-index multimode fibers are ideal for short-distance (up to several kilometers) data communications applications such as local area networks. For graded-index multimode fiber used in data communications, the standard NAs are 0.20 for 50/125 μm fiber and 0.275 for 62.5/125 μm fiber.

Defining NA

NA is a unitless quantity, shown in Fig. 2.1. It is derived from calculating the sine of the half angle (θ) of acceptance within the **cone** of light entering the fiber's core. Theoretical NA may be expressed by the equation $NA = (n_1^2 - n_2^2)/2$, where n_1 is the refractive index of the core and n_2 is the refractive index of the **cladding**. The refractive index of a material is defined as the ratio of the speed of light in a vacuum to the speed of light in that particular material.

---

critical *adj.* 关键的
performance *n.* 性能
specification *n.* 规格，说明书，规范
particular *adj.* 特殊的，特别的，独特的

Vertical Cavity Surface Emitting Laser
垂直腔面发射激光器

alignment *n.* 队列，顺序
tolerance *n.* 公差

cone *n.* [数、物]锥形物，圆锥体

cladding *n.* 覆层

Fig. 2.1 NA defines the maximum angle (the "cone of acceptance") at which light can be launched into a fiber

Get to Work

I. Matching

Directions: Match the words or expressions in the left column with the Chinese equivalents in the right column.

1. NA   A. 垂直腔面发射激光器
2. VCSEL   B. 数值孔径
3. LED   C. 激光
4. laser   D. 灯泡
5. lamp bulb   E. 发光二极管

II. Understanding Checking

Directions: Mark Y (for YES) if the statement agrees with the information given in the text; N (for NO) if the statement contradicts the information given in the text; NG (for NOT GIVEN) if the information is not given in the text.

(   )1. Numerical aperture indicates the maximum angle at which a particular fiber can accept the light that will be transmitted through it.

(   )2. Multimode fibers allow more modes of light to be transmitted, resulting in greater pulse spreading, or dispersion, and more bandwidth.

(   )3. NA is a unitless quantity.

(   )4. Theoretical NA may be expressed by the equation $NA = (n_1^2 - n_2^2)/2$, where n_1 is the refractive index of the core and n_2 is the refractive index of the cladding.

III. Passage Completion

Directions: Fill in each of the blanks with one of the words or expressions in the box, making changes if necessary.

| viewpoint | radiant | range | nature |
| phenomenon | definition | fear | sure |

Throughout human history, light has been something most of mankind has taken for granted. It is there throughout our lives for most of us, and (so we assume) will always be there in the familiar patterns we experienced as we grew up.

In the past, and in many countries even today, phenomena such as solar eclipses have been cause for great __1__, because they represent a break in that familiar pattern, cutting off the light from the sun for a while, and who could be __2__ if the sun would ever come back? Even in countries where an eclipse is an understood __3__, a solar eclipse is still an occasion for excitement and awe.

To gain any understanding of light itself, we need to step away from this mindset and examine light from a more scientific and objective __4__. Let's start with a dictionary __5__ of light, with some technical data included:

Light

The form of __6__ energy that stimulates the organs of sight, having for normal human vision wavelengths __7__ from about 3900 to 7700 ångstroms and traveling at a speed of about 186 300 miles per second.

$$1 \text{ ångstrom} = 10^{-8} \text{ cm } (0.00000001 \text{ cm}).$$

Of course, the above definition doesn't really tell us much. Before the speed of light and its wavelength in the electromagnetic spectrum were determined, the definition would have ended at the first comma, and that really would have told us nothing about the __8__ of light.

Ⅳ. Translation

Directions: Translate sentences from English into Chinese.

A new LED design employs a handy combination of light and phosphors to produce light whose color spectrum is not so different from that of sunlight.

LEDs convert electricity into light very efficiently, and are increasingly the preferred design for niche applications like traffic and automobile brake lights. To really make an impression in the lighting world, however, a device must be able to produce room light. And to do this one needs a softer, whiter, more color-balanced illumination.

2.4 Electromagnetic Radiation

Unit 2 Optics   45

Learning objectives:

In this section, you will:
- understand the complete range of electromagnetic radiation.
- find the reason why scientists said Marconi was wasting his time when he attempted to send radio waves from Cornwall to Newfoundland in 1902.
- comprehend why radio waves could be sent to bounce around the world.
- indicate where the different layers occur.
- understand that the presence of electrons and ions in the upper atmosphere makes it electrically conducting.

Warm-Up Activity

Describe the following pictures in one or two sentences.

Picture 1   Picture 2   Picture 3

Picture 4   Picture 5

Text

The complete range of electromagnetic radiation is made up of **gamma rays**, X-rays, **ultraviolet rays**, ordinary visible light, **infrared** (heat) **radiation**, and radio waves. All these are given out by the sun, but some of them do not reach us, as the **atmosphere behaves** like a filter. It lets through visible light and certain radio waves; most of the other radiation is absorbed. This is fortunate for us, because gamma rays, X-rays, and ultraviolet rays can harm living things. Enough ultraviolet rays get through for us to detect them—they are responsible for our getting **sunburned** when we expose ourselves too rapidly in sunbathing.

A well-known characteristic of light is that it travels in straight lines. The other electromagnetic radiations behave in a similar way. For this reason scientists thought **Marconi** was wasting his time when he attempted to send radio waves from **Cornwall** to **Newfoundland** in 1902. It seemed obvious that the **curvature** of the earth across the **Atlantic** would prevent this.

However, the experiment was successful, and once again it had been shown that experiments should always be made to check **predictions** from theory.

A new theory was soon developed to explain the new facts. This was that a layer existed in the upper atmosphere which could reflect radio waves. In this way they would be bounced round the world. This was called the **Heaviside layer**, after one of the scientists who predicted it; but as later work showed that there were several layers at different heights, they are now referred to by letters, and the general region in which they are present is called the **ionosphere**.

The D-layer occurs at a height of fifty to sixty miles and disappears during the night; the E-layer occurs at sixty to seventy **miles** and is weakened but still present at night. The F-layer extends from ninety miles upwards, and is present as two layers (F1 and F2) during the day. The layers are able to conduct electricity because the particles (atoms and molecules) of the atmospheric gases are **ionized**, that is to say, **split** into **ions** and electrons.

An atom of any substance consists of a positively-charged central body (the nucleus) surrounded by negatively-charged electrons, like **planets** around the sun. Each electron bears a single negative charge; and usually the **nucleus** carries a number of positive charges equal to

---

gamma ray 伽马射线
ultraviolet ray 紫外线
infrared *adj.* 红外线的
radiation *n.* 放射，发光，发热，辐射
atmosphere *n.* 大气，空气
behave *v.* 举动，表现
sunburned *adj.* 起晒斑的，晒黑的
Marconi *n.* 马可尼
Cornwall *n.* 康沃尔
Newfoundland *n.* 纽芬兰(岛)
curvature *n.* 曲率
Atlantic *n.* 大西洋
prediction *n.* 预言，预报

Heaviside layer *n.* 海维赛层，E 层

ionosphere *n.* 电离层

mile *n.* 英里

ionize *vt.* 使离子化 *vi.* 电离
split *v.* 分裂，分离
ion *n.* 离子
planet *n.* [天]行星
nucleus *n.* (原子)核

the number of electrons. Therefore the whole atom is neither positive nor negative. If one or more electrons are removed, the remainder of the atom must be positively charged. It is possible to add extra electrons to the atom, and then we get negative ions.

Electric currents flowing in wires are streams of electrons running through the wire. Electrons also flow in a television tube, where they are made to hit the screen, causing a flash of light. Ions can also act as a current of electricity in liquids or gases, and the presence of electrons and ions in the upper atmosphere makes it electrically conducting. In fact, if there is a wind in the ionosphere, it will be an electric current.

Get to Work

I. Matching

Directions: Match the words or expressions in the left column with the Chinese equivalents in the right column.

1. neutron   A. 原子
2. electron   B. 分子
3. ion   C. 离子
4. molecular   D. 中子
5. atom   E. 电子

II. Understanding Checking

Directions: Mark Y (for YES) if the statement agrees with the information given in the text; N (for NO) if the statement contradicts the information given in the text; NG (for NOT GIVEN) if the information is not given in the text.

(　　) 1. The atmosphere lets through visible light and certain waves.

(　　) 2. Light travels in straight lines.

(　　) 3. The curvature of the earth across the Atlantic would prevent Marconi's attempt in 1902.

(　　) 4. An atom of any substance consists of a positively-charged central body as well as negatively-charged electrons.

III. Passage Completion

Directions: Fill in each of the blanks with the corresponding expression in the box.

energy of the photon	the frequency of the wave
the index of refraction of the medium	Planck's constant
the velocity of light in open space	the wavelength of the wave

The basic transverse electromagnetic wave involves both a varying electric field and a varying magnetic field, appearing at right angles to each other and to the direction of travel of the wave. Note especially that the electric and magnetic fields are not in phase with each other, but are rather 90° out of phase. Most books portray these two components of the total wave as being in phase with each other, but I find myself disagreeing with that interpretation, based on three fundamental laws of physics.

The total energy in the waveform must remain constant at all times. Any deviation from this condition constitutes a violation of this law.

The energy in a single photon is given by the expression

$$E = h\nu = \frac{hc}{n\lambda}$$

In this expression:
$E=$ _____ ; $h=$ _____ ;
$\nu=$ _____ ; $c=$ _____ ;
$n=$ _____ ; $\lambda=$ _____ .

Ⅳ. Translation

Directions: Translate sentences from English into Chinese.

Electric currents flowing in wires are streams of electrons running through the wire. Electrons also flow in a television tube, where they are made to hit the screen, causing a flash of light. Ions can also act as a current of electricity in liquids or gases, and the presence of electrons and ions in the upper atmosphere makes it electrically conducting. In fact, if there is a wind in the ionosphere, it will be an electric current.

CAREER SPEAKING

Meeting Guests

Ⅰ. Warm-Up

Model

(A: Zhang Li   B: Mr. Mills)

A: Excuse me, are you Mr. Mills from New York?

B: Yes, I am.

A: Nice to meet you. I am Zhang Li, an **administrative** assistant of Wuhan Brightness

Optoelectronics Corporation.

B: Nice to meet you, too, Miss Zhang. I really appreciate your coming to meet me.

A: You're welcome. How was your flight?

B: Great. It was **uneventful**.

A: I'm very glad to hear that. Shall I take you to the hotel now? Our car is waiting over there.

B: Yes, that would be wonderful!

A: We've reserved one suit for you at **Shangri-La Hotel**, which is one of the best hotels in the city. I hope you have a good time there.

B: Thank you. You're really **considerate**.

A: My pleasure, Mr. Mills. In the evening, after the welcome party given by our manager, we'll take a car-ride round the city. At 8:30 tomorrow morning, I will drive to pick you up to our corporation. Is that all right?

B: All right. I'm here **at your disposal**.

Notes

administrative	adj. 行政的,管理的
uneventful	adj. 太平无事的
Shangri-La Hotel	香格里拉饭店
considerate	adj. 考虑周到的
at your disposal	听你安排

II. Matching

Directions: Please choose the corresponding sentences from the box to fill in the blanks, then play roles with your partner.

(A: Wang Yun   B: Zhang Wen)

A: _____

B: Yes, I am.

A: Hello! _____ Our manager has asked me to come and meet you.

B: _____

A: You're welcome. _____

B: Thank you.

A: Not at all. _____

B: Yes, very pleasant.

A: The waiting room is over there. Let's take a short rest. _____

> S1: Let me help you with your baggage.
> S2: Did you have a good trip?
> S3: Excuse me, are you Miss Zhang Wen from Sichuan?
> S4: And then I'll drive you to the hotel.
> S5: Nice to meet you. I'm Wang Yun, the secretary of Yichang Foreign Trade Company.
> S6: Nice to meet you, too, Miss Wang. Thank you for meeting me.

Ⅲ. Closed Conversation

Directions: Please fill in the blanks according to the Chinese, then play roles with your partners.

(A: Lily   B: Ben Green   C: Jerry)

A: Good morning, Mr. Green. _____

(请让我向你介绍我的同事杰利,他也是来接你的。)

B: How do you do, Jerry?

C: _____(你好,格林先生,欢迎来到武汉。)

B: Thank you. I prefer Ben, if that's OK.

C: Ben, how was the flight?

B: _____

(不太好。由于糟糕的天气,航班延误了好几个小时。)

C: After a long flight, you need a good rest.

A: Ben, if you don't mind, while you and Jerry get acquainted, _____

_____(我把车开过来,请稍等。)

B: All right. Jerry, when shall we get down to business?

C: _____

(我已经准备了日程表,请过目,好吗?)

B: OK. Let me see. Tomorrow afternoon we will have a meeting, right?

C: Yes. _____

(会议定于明天下午 2 点举行,我会在下午 1 点半来接您。)

B: OK. _____

(从日程表上看,我发现你们已经安排了两天的时间带我去市里到处游览。你们真是想得太周到了,谢谢。)

A: _____

(别客气。我们希望您在武汉过得愉快。车来了,我们走吧。)

IV. Semi-Open Conversation

Directions: Please fill in the blanks with your own words, then play roles with your partner.

(A: Susan   B: Jane)

A: Hi, Jane. _____

B: Me too. _____

A: My pleasure. _____

B: It was a pleasant trip, but I'm a little tired.

A: _____ Then you'll have a good rest.

B: Thank you. You're very thoughtful.

A: Don't mention it. _____ Let's go.

V. Live Show

Challenge 1

Directions: Mr. Zhang is asked to meet Mr. Smith from San Francisco at the airport. They greet each other and then Mr. Zhang drives Mr. Smith to the hotel. Please act it out with your partner or partners. The model and the useful expressions below are just for your reference.

Challenge 2

Directions: David, together with his colleague Miss Wu, on behalf of the manager of Happiness Import & Export Company, is asked to pick up John. David greets his old friend John and introduces Miss Wu to him. They three talk about the flight and the arrangement of business. Please act it out with your partner or partners. The model and the useful expressions below are just for your reference.

Model

(A: Henry Jones   B: Sally   C: Liu Xin)

B: Good morning, Henry. Delighted to see you again.

A: Me too, Sally. It's been quite a while, hasn't it?

B: Right. It's been almost one year since I saw you last time, but you haven't changed a bit.

A: Thank you. Neither have you.

B: I would like to introduce Mr. Liu Xin to you. Mr. Liu is the manager of our **import division**. Mr. Liu, this is Henry Jones from Good Luck Import and Export Co. Ltd.

C: Glad to meet you, Mr. Jones. We've been looking forward to seeing you.

A: Glad to meet you, too. Mr. Liu. Thank you for your coming to meet me.

C: It's my pleasure. Did you enjoy your trip?

A: Yes, it was a pleasant trip. But I am feeling a little **jet lag**.

C: If you take a rest, I'm sure you'll be fine tomorrow, Mr. Jones.

A: Henry, please. "Mr. Jones" makes me sound ancient.

C: OK, Henry. You can also call me Peter, my English name.

B: Henry, please give me your baggage **check**. I'll go and **claim the baggage** for you.

A: Here you are. Thanks.

B: You're welcome. I'll be back soon.

C: How do you like the weather here, Henry?

A: It's a lovely day. The weather here is quite similar to that in our city.

C: As far as I know, you have been in Wuhan for several times. Do you get accustomed to it?

A: Yes. I think it's all right. I like Wuhan. It's a **fantastic** place.

C: You must have visited many **scenic spots and historical sites**.

A: Frankly speaking, I am always too busy to spare any time for a tour.

C: If you like, we'll set up a **sightseeing** program for you.

A: That's great. It's very kind of you, Peter.

C: Please think nothing of it. Sally will take care of you during your stay in Wuhan. Don't hesitate to call her if you need any help. I hope you **feel at home** here.

A: Thank you very much. Sally is my good friend. She is really warm-hearted and thoughtful.

C: I'm glad to hear that. Oh, here she comes. We have a car waiting outside to take you to your hotel. Shall we go now?

A: Yes, let's go.

Notes

import division	进口部
jet lag	飞行时差反应
check	*n.* 用以识别的票证或纸片
claim the baggage	领行李
fantastic	*adj.* 极好的
scenic spots and historical sites	名胜古迹
sightseeing	*n.* 观光
feel at home	感到像在家一样舒适

Useful expressions

1. 见面寒暄

Hi, how's your business?

Delighted to see you again.

It's a pleasure to meet you at last.

I've heard so much about you.

It's very kind of you to come here to meet me.

2. 相互介绍

Allow me to introduce myself.

May I introduce Tom to you?

I'd like you to meet my colleague Mary.

This is Linda from England.

3. 旅途情况

Did you have a good trip?

How was your flight?

Yes, it was a very smooth flight.

Not so good. There was a thunderstorm so the flight was delayed for several hours.

This one was uneventful, except for a little turbulence here and there.

4. 去宾馆的路上

I hope you'll enjoy your stay here.

I've made a reservation at the hotel you used last time.

I've booked a room for you at Hilton Hotel. Single, for 9 days.

I'll have the car brought here, so please wait a moment.

5. 日程安排

It's just the matter of the schedule, that is, if it is convenient for you right now. We'll leave some evenings free, if it is all right with you.

The meeting is scheduled for 3:30 this afternoon. I will pick you up at 3:00 p.m. I have arranged your schedule. I hope it's suitable.

We've arranged one afternoon in the time schedule to show you around the city.

WRITING

Letter of Thanks

在人际交往中,对别人的帮助、鼓励、馈赠的礼物、提供的服务等表示真诚的感谢是很有必要的,既能体现自身的道德素质,又能拉近人与人之间的关系。除了当面道谢和电话道谢外,用别致的卡片或短信也能表达自己的心意。在写感谢信时要特别注意以下几点。

（1）开门见山，言简意赅，说明感谢的原因。

（2）措辞自然，态度诚恳，不可夸大其词、不切实际。

（3）表达自己的良好祝愿。

Sample 1

May 22, 2011

Dear Mr. Smith,

I am writing this letter to thank you for the wonderful time I spent during my stay in Wuhan. I was **overwhelmed** by your warm hospitality which made me feel at home, and I was impressed by the visit to your company which was very informative and **inspiring**.

I would like to express my best wishes for the success in your career.

Yours sincerely,

Bill Lawrence

Notes

overwhelm　　　　　　　　　　v. （感情或感觉）充溢，难以禁受

inspiring　　　　　　　　　　　adj. 令人鼓舞的，启发灵感的

Sample 2

Dec. 19, 2011

Dear Miss Wang,

Thank you for your letter of Dec. 16, enclosing the catalogues and price lists, which fascinate us very much. I find the products with new designs and favorable prices gain an edge over your competitors. We really appreciate your sincerity and look forward to establishing trade relations with your company.

Best wishes.

Yours faithfully,

Nancy Smith

Useful expressions

1. Many thanks for your …

2. Please accept my sincere appreciation for …

3. I am writing this letter to thank you for …

4. Thank you more than I can say.

5. I take this opportunity to express to you my deep appreciation for …

Challenge

Your colleague Miss Zhang sent you a gift for your wedding. You write a letter to express your thanks.

EXTENSIVE READING

2.5 What Is Light

Throughout human history, light has been something most of mankind has taken for granted. It is there throughout our lives for most of us, and (so we assume) will always be there in the familiar patterns we experienced as we grew up.

In the past, and in many countries even today, phenomena such as solar eclipses have been cause for great fear, because they represent a break in that familiar pattern, cutting off the light from the sun for a while, and who could be sure if the sun would ever come back? Even in countries where an eclipse is an understood phenomenon, a solar eclipse is still an occasion for excitement and awe.

To gain any understanding of light itself, we need to step away from this mindset and examine light from a more scientific and objective viewpoint. Let's start with a dictionary definition of light, with some technical data included:

Light

The form of radiant energy that stimulates the organs of sight, having for normal human vision wavelengths ranging from about 3900 to 7700 ångstroms and traveling at a speed of about 186 300 miles per second.

$$1 \text{ ångstrom} = 10^{-8} \text{ cm } (0.00000001 \text{ cm}).$$

Of course, the above definition doesn't really tell us much. Before the speed of light and its wavelength in the electromagnetic spectrum were determined, the definition would have ended at the first comma, and that really would have told us nothing about the nature of light.

So, rather than look at more definitions, let's move on to explore some of the basic properties of light as we know them today, so we can better understand not only how light will behave, but also something of why it behaves as it does. This will give us a chance to predict how light may behave under various circumstances and conditions.

Exercise: Can you tell us what light is in your own words?

2.6 Convex Lens

The most commonly-seen type of lens is the convex lens. This type of lens is often used for close examination of small objects, such as rare stamps or coins. Children often use such a lens to concentrate sunlight to burn small pinholes in pieces of paper.

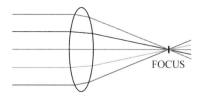

That result by itself shows the power of concentrated light from the sun. But there must be more to it than that. Let's see if we can define the behavior of lenses a bit more specifically.

The figure to the right shows a double convex lens with several rays of light approaching from its left. We show each ray as a different color here, simply to more easily follow each ray's progress. We will assume that the lens is made of glass with a nominal index of refraction of 1.50.

The rays are parallel as they approach the lens. As each ray reaches the glass surface, it refracts according to the effective angle of incidence at that point of the lens. (See the pages on refraction for the definitions and descriptions of these terms.) Since the surface is curved, different rays of light will refract to different degrees; the outermost rays will refract the most.

As the light rays exit the glass, they once again encounter a curved surface, and refract again. This further bends the rays of light towards the centerline of the lens (which coincides with the green light ray in the figure).

Exercise: Translate the following into Chinese.
convex lens   nominal index of refraction   effective angle   degree   ray

TRANSLATION

Translation of Words and Phrases

词语的翻译

由于中英两种语言的内在结构不同,以及人们的思维习惯、表达方式存在差异,因此在翻译过程中要考虑目的语读者的阅读习惯,如实地传达原文的思想内容,使译文通顺自然。在翻译词语时,通常有以下几种方法。

1. 直接翻译法

保持原文的形式和内容,根据词语字面意思翻译出来。

Three types of printers are available: wire printers, ink jet printers and laser printers.

有三种类型的打印机:针式打印机、喷墨打印机和激光打印机。

2. 转换翻译法

为了使译文更符合汉语的表达习惯,有时采用直译会觉得别扭、不自然,这就需要进行词类转换。

1) 转换为动词

Attempts have been made to rename this machine.

人们一直在**尝试**给这台机器重新命名。
(名词 attempt 转换为动词"尝试")
2) 转换为名词
The average man who uses a laser printer could not explain how it **works**.
普通人只会使用激光打印机,却无法解释它的**工作原理**。
(动词 work 转换为名词"工作原理")
3) 转换为副词
Today's lasers are producing parts faster and with **higher quality**.
现代激光技术可以更快**更好地**制造各种部件。
(名词短语 higher quality 转换为副词"更好地")
4) 转换为形容词
The ozonosphere is of great **importance** to mankind.
臭氧层对人类是非常**重要的**。
(名词 importance 转换为形容词"重要的")

3. 引申翻译法

有时采用直译法无法表达词语内在包含的意思,这时就需要根据原文的意思,按照汉语的表达习惯,对词义进行引申。
This subject is still in its **infancy**.
这门学科仍处于**发展初期**。
(infancy 引申为"发展初期")

4. 增词翻译法

在译文中增加英语原文中没有的词,使表达的意思更准确、更完整。
A drilled hole can be made accurate and smooth by a reamer.
钻好的孔可用绞刀加工,使其**尺寸**精确、**表面**光滑。
(增译"尺寸"和"表面")

5. 省略翻译法

在英语中有很多词,如冠词、介词、连词、代词,甚至是动词等,在翻译时可以省略,这样使译文更自然流畅,避免生硬难懂。
A wave model **is considered** as a good model of the way in which light behaves.
波动模型是说明光的性能的一个很好的模型。
(省译 is considered)

6. 重复翻译法

汉语注重形式,讲究文采,通常会重复某个词语以加强语气,平衡结构,使语言富于表现力。
Fine-blanking press can be **driven** both mechanically and hydraulically.
精密冲切机既可以机械**传动**,也可液压**传动**。
(重复 driven)

Challenge

Translate the following sentences into Chinese and pay special attention to the parts in bold.

1. These are modern materials having greater uniformity **and** higher optical sensitivity.
2. It takes years to gain a **mastery** of the new technology.
3. The high intensity of a laser beam makes it a **convenient** drill.
4. The laser is shown **briefly** in the following chapter.
5. The **micrometer** is marked in hundredths of a millimeter.

TASK

Electromagnetic Radiation

Target	Understand and express professional knowledge.				
Requirements	Make an English presentation according to lesson 2.4. Produce a micro-video to express professional knowledge. Add accurate Chinese and English subtitles to the video.				
Contents	Electromagnetic Radiation: Complete range of electromagnetic radiation Experiment Ionosphere Conducting				
Group: _____	**Project**	**Name**	**Software**	**Time**	**Score**
	Material collection				
	Script				
	PPT				
	Reader				
	Subtitle				
	MP4				
Operating environment	XP/Win7/Win8/ Mobile phone				
Product features					

Script reference

标题	专业内容(讲稿与中、英文字幕)	PPT 中需显示的文字	表现形式	素 材
每个知识点的要点或者小节名字	做报告时的英文讲稿 视频中的中、英文字幕	需要特别提醒的文字	讲解到某个地方或某句话时,出现的图片、动画或者文字	图片或者视频
Contents	**Electromagnetic Radiation** Contents: Complete range of electromagnetic radiation Experiment Ionosphere Conducting 电磁辐射 内容: 整个电磁辐射范围 实验 电离层 导电	**Electromagnetic Radiation**: Complete range of electromagnetic radiation Experiment Ionosphere Conducting	文字	
1. Complete range of electromagnetic radiation 整个电磁辐射范围	The complete range of electromagnetic radiation is made up of **gamma rays**, X-rays, **ultra-violet-rays**, ordinary visible light, **infra-red** (heat) **radiation**, and radio waves. 整个电磁辐射范围由 r 射线、X 射线、紫外线、普通可见光线、红外辐射和无线电波组成。	Complete range of electromagnetic radiation: **gamma rays**, X-rays, **ultra-violet-rays**, ordinary visible light, **infra-red** (heat) **radiation**, and radio waves.	图解	图 1 电磁波谱图

续表

标题	专业内容(讲稿与中、英文字幕)	PPT 中需显示的文字	表现形式	素 材
2. Experiment 实验	All these are given out by the sun, but some of them do not reach us, as the **atmosphere behaves** like a filter. 所有这些射线都由太阳发射出来，但是其中一些射线并不能到达我们这里，因为大气层起到了过滤作用。		图解	图 2 大气过滤
	A well-known characteristic of light is that it travels in straight lines. The other electromagnetic radiations behave in a similar way. 光的一个尽人皆知的特性是直线传播。其他电磁辐射也一样。	Experiment	文字	
	For this reason, scientists thought **Marconi** was wasting his time when he attempted to send radio waves from Cornwall to **Newfoundland** in 1902. 因此，科学家们当时认为，马可尼在 1902 年想把无线电波由康沃尔发送到纽芬兰是浪费时间。	**Marconi** attempted to send radio waves from Cornwall to **Newfoundland** in 1902.	图解	图 3 科学家马可尼
	It seemed obvious that the **curvature** of the earth across the **Atlantic** would prevent this. 横跨大西洋的地球弧度会使这个想法不能实现，这似乎是显而易见的。	It seemed obvious that the **curvature** of the earth across the **Atlantic** would prevent this.		图 4 地球曲率
	However, the experiment was successful, and once again it had been shown that experiments should always be made to check **predictions** from theory. 然而实验成功了，它又一次说明，从理论上推出的预见是需要通过实验来检验的。	However, the experiment was successful.		

续表

标题	专业内容（讲稿与中、英文字幕）	PPT 中需显示的文字	表现形式	素 材
3. Ionosphere 电离层	A new theory was soon developed to explain the new facts; this was that a layer existed in the upper atmosphere which could reflect radio waves. In this way they would be bounced round the world. 为了解释这种新的事实，很快又提出了一种新理论，即在大气层的上部，有一个能反射无线电波的层次。这样，无线电波可以绕地球多次反射。	A new theory was soon developed to explain the new facts; this was that a layer existed in the upper atmosphere which could reflect radio waves. In this way they would be bounced round the world.		图 5 无线电波多次反射图
	This was called the **Heaviside** layer, after one of the scientists who predicted it; but as later work showed that there were several layers at different heights, they are now referred to by letters, and the general region in which they are present is called the **ionosphere**. 这一层叫海维赛层，是以一位对此做出预言的科学家的名字命名的。但是由于后来的研究表明，在不同的高度分别有好多层。现在，各层都用字母来表示，而它们所处的总区域叫作电离层。	There were several layers at different heights. They are now referred to by letters, and the general region in which they are present is called the **ionosphere**.		图 6 电离层
D-layer	The D-layer occurs at a height of fifty to sixty miles and disappears during the night. D 电离层位于 50 至 60 英里的高度，夜间便消失。	The D-layer occurs at a height of fifty to sixty miles and disappears during the night.	示意图，图解强调 D 层	图 6 电离层
E-layer	The E-layer occurs at sixty to seventy **miles** and is weakened but still present at night. E 电离层位于 60 至 70 英里的高度，夜间减弱，但一直存在。	The E-layer occurs at sixty to seventy **miles** and is weakened but still present at night.	示意图，图解强调 E 层	图 6 电离层

续表

标题	专业内容（讲稿与中、英文字幕）	PPT中需显示的文字	表现形式	素 材
F-layer	The F-layer extends from ninety miles upwards, and is present as two layers (F1 and F2) during the day. F 电离层从 90 英里向上延伸，白昼是两层(F1 和 F2)。	**The F-layer** extends from ninety miles upwards, and is present as two layers (F1 and F2) during the day.	示意图，图解强调 F 层	图 6 电离层
4. The layers are able to conduct electricity 各层均能导电	The layers are able to conduct electricity because the particles (atoms and molecules) of the atmospheric gases are **ionized**, that is to say, **split** into **ions** and electrons. 各层均能导电，因为大气粒子（原子和分子）被电离，也就是说被分裂为离子和电子。	The layers are able to conduct electricity because the particles (atoms and molecules) of the atmospheric gases are **ionized**, that is to say, **split** into **ions** and electrons.		图 7 原子模型
	The presence of electrons and ions in the upper atmosphere makes it electrically conducting. In fact, if there is a wind in the ionosphere, it will be an electric current. 大气上层存在电子和离子，使其具有导电性。事实上，如果电离层中有风的话，那大气就是电流。			图 7 原子模型

UNIQUE ENGLISH

Incredible Sentences

1. I think that student wrote on the blackboard was wrong.
2. I know. You know. I know that you know. I know that you know that I know.

3. 2B or not 2B, that is a ?
4. To China for china, China with china, dinner on china.
5. The quick brown fox jumps over a lazy dog.
6. Was it a bar or a bat I saw?

Unit 3
Laser

In this unit, you are going to learn the following contents:

READING IN DEPTH

3.1　This Is a Laser

3.2　Applications of Lasers

3.3　Laser Construction

3.4　Internet and Lasers

CAREER SPEAKING

Introducing Enterprises

WRITING

Letter of Congratulation

EXTENDSIVE READING

3.5　A Prospectus about Laser Machines

3.6　Using Lasers

TRANSLATION

Translation of Noun Clauses

TASK

Laser Construction

UNIQUE ENGLISH

Funny Chinglish

READING IN DEPTH

3.1 This Is a Laser

Learning objectives:

In this module, you will:
● understand the meaning of the word "laser".
● learn about applications of lasers.
● master differences between lasers and other light.

Warm-Up Activity

Describe the following pictures.

Words and phrases for reference: play guitar, a fighter equipped with a laser weapon, a frog, open an umbrella, laser cutting, pull a long face

Picture 1

Picture 2

Picture 3

Picture 4   Picture 5

Text

Most people know the word "laser", but do they know what it really is? What's the difference between ordinary light and lasers, and what does laser really **stand for**? Let's start with the last question.

Laser is an **acronym**, which is a word made up of **initial** letters. You could use the complete name: light **amplification** by **stimulated emission** of radiation, but that's a bit **awkward**. Let's **keep to** laser.

Almost everyone probably knows that the police use a laser when they measure speed. At least many drivers that have **exceeded** the **speed limit** know about it, but how many know that you also use lasers several times during an ordinary day? You'll find them in **CD** players, laser printers and much, much more.

You often find lasers in action movies where the hero has to escape the laser beams when he's trying to solve a **thrilling** problem. The power contained in a laser is both **fascinating** and frightening.

---

stand for 代表,象征

acronym *n.* 首字母缩略词
initial *adj.* 最初的,词首的
amplification *n.* 扩大
stimulate *vt.* 刺激,激励
emission *n.* 散发,发射
awkward *adj.* 难使用的,笨拙的
keep to (使)谈话不离题
exceed *vt.* 超越
speed limit 最高速度限制,速度极限
CD (compact disc) 光盘
thrilling *adj.* 惊险的,紧张的
fascinating *adj.* 迷人的,着魔的

Now, let's turn to the difference between laser light and other light.

Light is really an **electromagnetic** wave. Each wave has brightness and color, and **vibrates** at a certain angle, so-called **polarization**. This is also true for laser light but it is more **parallel** than any other light source. Every part of the beam has (almost) the exact same direction and the beam will therefore **diverge** very little. With a good laser an object at a distance of 1 km (0.6 mile) can be **illuminated** with a dot about 60 **mm** (2.3 inches) in **radius**.

As it is so parallel it can also be focused to very small **diameters** where the **concentration** of light energy becomes so great that you can cut, **drill** or turn with the beam. It also makes it possible to illuminate and examine very tiny details. It is the property that is used in **surgical appliances** and in CD players. It can also be made very **monochromic**, so that just one light **wavelength** is present. This is not the case with ordinary light sources. White light contains all the colors in the **spectrum**, but even a colored light, such as a red LED (light emitting diode) contains a continuous **interval** of red wavelengths.

On the other hand, laser emissions are not usually very strong when it comes to energy content. A very powerful laser of the kind that is used in a laser show does not **give off** more light than an ordinary **streetlight**; the difference is in how parallel it is.

electromagnetic *adj.* 电磁的
vibrate *v.* （使）振动，（使）颤动
polarization *n.* ［物］偏振(现象)，分化
parallel *adj. &.n. & v.* 平行(的)
diverge *v.* 分叉，分歧
illuminate *vt.* 照明，说明
mm *abbr.* millimeter 毫米
radius *n.* 半径
diameter *n.* 直径
concentration *n.* 集中，专心，浓缩，浓度
drill *v.* 训练，钻孔
surgical *adj.* 外科手术的
appliance *n.* 工具，设备
monochromic *adj.* 单色的
wavelength *n.* ［物］波长
spectrum *n.* 光谱，型谱，频谱
interval *n.* 间隔，距离，幕间休息

give off 发出(蒸汽、光等)，长出(枝、杈等)
streetlight *n.* 街灯

Get to Work

I. Matching

Directions: Match the words or expressions in the left column with the Chinese equivalents in the right column.

1. air-proof   A. 固定光束
2. beam axis   B. 激光探测器
3. optical guiding system   C. 远红外区
4. fixed beam   D. 光学头
5. perfect lens   E. 密封的
6. laser detector   F. 玻璃棒
7. optical path   G. 射束轴
8. optical head   H. 理想透镜
9. far-infrared region   I. 光程
10. glass rod

J. 光制导系统

Ⅱ. **Understanding Checking**

Directions: Mark Y (for YES) if the statement agrees with the information given in the text; N (for NO) if the statement contradicts the information given in the text; NG (for NOT GIVEN) if the information is not given in the text.

( )1. Since "light amplification by stimulated emission of radiation" seems to go beyond our understanding, let's give it up.

( )2. Lasers find their application in measuring car speed, CD players, printers and so on.

( )3. The hero in action movies often comes across the laser beams.

( )4. Compared with other light, a laser diverges little.

( )5. Light given off by a laser is more than that of an ordinary streetlight.

Ⅲ. **Passage Completion**

Directions: Fill in each of the blanks with one of the words or expressions in the box, making changes if necessary.

introduce	develop	do	categorize	tool
trend	supplier	settle	overview	apply

The TRUMPF Group is one of the world's leading corporations in production technology, with a total of approximately 6 500 employees at 43 locations in 23 countries. Our innovations set the __1__: we are the leader for industrial laser technology and laser systems in the global market, and are among the largest machine tool __2__ worldwide, because we never have __3__ for second-rate solutions. This __4__ not only to our products, but also to the names we give them.

We __5__ the names of our machines, lasers, programming systems, tools and accessories parallel with our product range in the past. As our product range grew over the years, so __6__ the number of product names.

For that reason, we __7__ a new name concept now. This will initially apply to our two largest business fields: machine __8__ and laser technology.

These new names are intended to provide an improved __9__ of our product program structure. Machines can now __10__ easily according to their new distinctive names.

Ⅳ. **Translation**

Directions: Complete the sentences by translating into English according to the Chinese given in brackets.

1. Before long the laser's distinctive qualities —_____
（能生成高强度、单一波长的窄光束）—were recognized.

2. Lasers are everywhere: from research laboratories at the cutting edge of quantum physics to _____
_____（诊所、超市收银处和电话网络）.

3. He analyzed the spectrum of light emitted and found _____
（其频率范围明显缩小）.

4. Theodore Maiman _____
（使第一台激光器成功运行）on 16 May 1960 at the Hughes Research Laboratory in California.

5. The scientist did not report having seen a bright beam of light, _____
_____（这是人们所预期的激光器特征）.

Ⅴ. **Sentence Remaking**

Directions: Simulate the following sentence patterns according to an example provided from the text.

1. ... is/are ... , which is/are ... made up ...

E.g. Laser is an acronym, which is a word made up of initial letters.

_____ is/are _____ , which is/are _____ made up of _____ .

2. ... find ... in ... where ...

E.g. You often find lasers in action movies where the hero has to escape the laser beams.

_____ find/finds _____ in _____ where _____ .

3. ... make it possible to ...

E.g. It also makes it possible to illuminate and examine very tiny details.

_____ makes/make it possible to _____ .

4. It is ... that ...

E.g. It is the property that is used in surgical appliances and in CD players.

It is _____ that _____ .

5. On the other hand, ... is/are ... when it comes to ...

E.g. On the other hand, laser emissions are not usually very strong when it comes to energy content.

On the other hand, _____ are/is useful when it comes to _____ .

Ⅵ. Para. Translation

In 1957, Charles Hard Townes and Arthur Leonard Schawlow, then at Bell Labs, began a serious study of the infrared maser (红外线微波激射器). As ideas were developed, frequencies (频率) were abandoned with focus on visible light instead. The concept was originally known as an "optical maser" (光量子放大器). Bell Labs filed a patent application (专利申请) for their proposed optical maser a year later. Schawlow and Townes sent a manuscript of their theoretical calculations to *Physical Review*, which published their paper that year.

Ⅶ. Mini Imitative Writing

Directions: The following paragraph originates from the text. Please imitate it and write another one. The following topic may be just for reference: MASER, LED, VCD, CCD, MOS, etc.

Laser is an acronym, which is a word made up of initial letters. You could use the complete name: light amplification by stimulated emission of radiation, but that's a bit awkward. Let's keep to laser.

3.2 Applications of Lasers

Learning objectives:

In this module, you will:
- understand industrial applications of lasers.
- master laser application in environment and communications.
- learn about lasers' role in research.

Warm-Up Activity

Describe the following pictures.

Phrases for reference: find one's application in; be used in; apply sth. to; put to use; find utilization in; application of sth. to

An example for Picture 1: We can find laser application in joining.

joining
Picture 1

cutting
Picture 2

surface treatment
Picture 3

micro processing
Picture 4

marking
Picture 5

surgery
Picture 6

Text

No one knew what a laser would be used for when it was discovered. "A solution waiting for a problem" is a famous saying from the early days of lasers. Sometimes it takes quite a long time before a discovery can be used in a **tangible** way. Yes, it might take 30 years or more before someone realizes that a discovery is really important. Today, lasers are used in several areas, such as research, communication, industry, medicine, and **environmental** care.

tangible *adj.* 现实的

environmental *adj.*
周围的,环境的

Industrial Applications of Lasers

Today, lasers can be found in a broad range of applications within industry, where they can be used for such things as pointing and measuring. In the manufacturing industry, lasers are used to measure the ball **cylindricity** in **bearings** by observing the **dispersion** of a laser beam when reflected on the ball. Yet another example is to measure the shadow of a **steel band** with the help of a laser light to find out the thickness of the band.

Within the **pulp** mill industry the concentration of **lye** is measured by observing how the laser beam **refracts** in it.

A laser also works as a **spirit level** and can be used to indicate a flat surface by just **sweeping** the laser beam along the surface. This is, for instance, used when making walls at **building sites**. In the **mining** industry, lasers are used to **point out** the drilling direction.

Environmental Applications

Laser technologies have also been used within environmental areas. One example is the ability to determine **from a distance** the environmental **toxins** in a column of smoke. Other examples are being able to predict and measure the existence of **photochemical smog** and **ozone**, both at ground level where it isn't wanted and in the upper **layers** of the atmosphere where it is needed. Lasers are also used to **supervise** wastewater **purification**.

Communications

A laser works as a light source in all fiber optics in use. It has greater **bandwidth** (potentially 100,000 times greater) than an ordinary **copper cable**.

cylindricity *n.* 圆柱度
bearing *n.* 轴承
dispersion *n.* 传播，散射
steel band 钢带

pulp *n.* 果肉，纸浆
lye *n.* 碱液
refract *vt.* 使折射，折射
spirit level 水平仪
sweep *v.* 扫，掠过
building site 建筑工地
mine *vt.* 开矿
point out 指出，指示
from a distance 从远方
toxin *n.* 毒素
photochemical *adj.* 光化学的
smog *n.* 烟雾
ozone *n.* 臭氧
layer *n.* 层
supervise *v.* 监督，管理，指导
purification *n.* 净化
bandwidth *n.* 带宽
copper cable 铜质电缆

It is **insensitive** to **interference** from external electrical and **magnetic** fields. **Crosstalk** (hearing someone else's phone call) is of rare **occurrence.**

Fiber optics is used increasingly often in data and telecommunications around the world.

Research

Lasers have become one of the most important tools to researchers within physics, chemistry, biology and medicine throughout the world and are used to:

— register **ultra-fast chemical processes** such as the **bonding** between **atoms** to form **molecules**;

— study the process when cells **split**, or a virus enters into a cell.

The full meaning of this research won't find its answer today but will be shown in the future.

insensitive *adj.* 不敏感的,麻木的
interference *n.* 干涉
magnetic *adj.* 磁的,有磁性的
crosstalk *n.* [电子学]串话干扰,串台
occurrence *n.* 发生,出现

ultra-fast chemical process 超快化学过程
bonding *n.* 连接,结合
atom *n.* 原子
molecule *n.* 分子
split *v.* 分裂

Get to Work

Ⅰ. Matching

Directions: *Match the words or expressions in the left column with the Chinese equivalents in the right column.*

1. optical flat
2. absorbing crystal
3. digital hologram
4. high vacuum
5. false image
6. maximum scale
7. arc discharge
8. heat diffusion
9. penetration frequency
10. acoustic signal

A. 高真空
B. 穿透频率
C. 最大刻度
D. 热扩散
E. 光学平面
F. 数字全息图
G. 误像
H. 声频信号
I. 吸收晶体
J. 电弧放电

Ⅱ. Understanding Checking

Directions: *Fill in the missing parts of the short passage with no more than 3 words based on the above text.*

The invention of a laser was a great event in the history, but _____ had passed before its wide use was found.

Lasers are being put in use in industry, e. g. they are used to _____ the ball cylindricity and the shadow of a steel band.

Lasers also are applied to environmental protection. For instance, they can monitor _____ purification.

Application of lasers into communications is one example of their great usage. As they are _____ to negative effects from external electrical and magnetic fields, they are effective in telecommunications.

Research is a field hoping for laser application. Lasers have proved their value by means of recording very fast _____.

Ⅲ. Passage Completion

Directions: Fill in each of the blanks with one of the words or expressions in the box, making changes if necessary.

| beam | achievement | maser | ruby rod |
| fame | breakthrough | win | electrical engineering |

The inventor of the first laser, Theodore Maiman, died on May 5, at the age of 79. Maiman studied __1__ at the University of Colorado and Stanford with Willis Lamb, who __2__ the Nobel Prize for physics in 1955. Maiman earned his PhD in 1955 — two years after Charles Townes invented the first __3__, the laser's predecessor generating amplified microwave __4__. In May 1960, Maiman made a __5__ in the development of the laser. He was able to generate pulses of coherent light by using a __6__. In 1984, Maiman received the prestigious Wolf Prize for his __7__. In the same year he became a member of the United States National Inventors Hall of __8__.

Ⅳ. Translation

Directions: Complete the sentences by translating into English according to the Chinese given in brackets.

1. _____(激光用于医疗) to improve precision work like surgery.

2. The coherency (一致性), high monochromaticity (高单色性), and _____ _____(能产生高能量的特性) are all properties of laser.

3. By _____(精心设计激光器件), the purity of the laser light can be improved more.

4. In 1917, Albert Einstein in his paper *On the Quantum Theory of Radiation*(《论辐射的量子性》), _____(为激光的发明奠定基础).

5. In 1939, Valentin A. Fabrikant predicted the use of stimulated emission to _____ _____(放大短波).

Ⅴ. Sentence Remaking

Directions: Simulate the following sentence patterns according to an example provided from the text.

1. It takes a long time before …

E. g. It takes quite a long time before a discovery can be used in a tangible way.

It takes a long time before _____ .

2. In … , … is/are used to …

E. g. In the mining industry, lasers are used to point out the drilling direction.

In _____ , _____ is/are used to _____ .

3. One example is … to …

E. g. One example is the ability to determine from a distance the environmental toxins in a column of smoke.

One example is _____ to _____ .

4. … is/are insensitive to … from …

E. g. It is insensitive to interference from external electrical and magnetic fields.

_____ is/are insensitive to _____ from _____ .

5. … won't … but …

E. g. The full meaning of this research won't find its answer today but will be shown in the future.

_____ won't _____ but _____ .

Ⅵ. Para. Translation

A technique that has had recent success is laser cooling. This involves atom trapping, a method where a number of atoms are confined in a specially shaped arrangement of electric and magnetic fields. Shining particular wavelengths of laser light at the ions or atoms slows them down, thus cooling them. As this process is continued, they all are slowed and have the same energy level, forming an unusual arrangement of matter known as a Bose-Einstein condensate.

Ⅶ. Mini Imitative Writing

Directions: The following paragraph originates from the text. Please imitate it and write another one.

No one knew what a laser would be used for when it was discovered. "A solution waiting for a problem" is a famous saying from the early days of lasers. Sometimes it takes quite a long time before a discovery can be used in a tangible way.

3.3 Laser Construction

> **Learning objectives:**
>
> In this module, you will:
> - understand the main parts of a laser.
> - learn about the types of gain media.
> - master the functions of the pump source and the optical resonator.

Warm-Up Activity

Write an individual signature（个性签名）for each of the following machines as if they are personified. You may discuss with your partner with the help of Picture 1 as an example.

Picture 1: Solid-State Laser

Picture 2: Gas Laser

Individual signature: "My predecessor was a top one in the working guys, so are his offspring."

Individual signature: _____
_____
_____.

Picture 3: Excimer Laser

Individual signature: _____
_____
_____.

Picture 4: Dye Laser

Individual signature: _____
_____
_____.

Text

A laser is constructed from three principal parts: An energy source (usually referred to as the pump or pump source), a **gain medium** or laser medium, and two or more mirrors that form an **optical resonator**.

Pump Source

The pump source is the part that provides energy to the laser system. Examples of pump sources include electrical **discharges**, **flashlamps**, **arc lamps**, light from another laser, chemical reactions and even explosive devices. The type of pump source used principally depends on the gain medium, and this also determines how the energy is transmitted to the medium. A **helium-neon** (HeNe) laser uses an electrical discharge in the helium-neon gas mixture, a **Nd:YAG** laser uses either light focused from a **xenon flash lamp** or **diode lasers**, and **excimer** lasers use a chemical reaction.

Gain Medium

The gain medium is the major determining factor of the wavelength of operation, and other properties of the laser. The gain medium is excited by the pump source to produce a **population inversion**,

gain medium 增益介质
optical resonator 光学谐振腔

discharge n. 放电
flashlamp n. 闪光灯
arc lamp 弧光灯
helium n. 氦
neon n. 氖
Nd:YAG (neodymium-doped yttrium aluminum garnet) n. 掺钕的钇铝石榴石
xenon flash lamp 氙气闪光灯
diode laser 二极管激光器
excimer n. 受激准分子
population inversion 粒子数反转

and it is in the gain medium that **spontaneous** and **stimulated emission** of **photons** takes place, leading to the phenomenon of optical gain, or amplification.

Examples of different gain media include:

Liquids, such as **dye**. The exact chemical configuration of the dye molecules determines the operation wavelength of the dye laser.

Gases, such as **carbon dioxide** and mixtures such as helium-neon. These lasers are often **pumped** by electrical discharge.

Solids, such as **crystals** and glasses. The solid host materials are usually **doped** with an **impurity**.

Semiconductors, a type of solid, in which the movement of electrons between material with differing **dopant** levels can cause laser action.

Optical Resonator

The optical resonator, or optical **cavity**, in its simplest form is two parallel mirrors placed around the gain medium which provide **feedback** of the light. The mirrors are given optical **coatings** which determine their reflective properties. Typically one will be a **high reflector**, and the other will be a **partial reflector**. The latter is called the output **coupler**, because it allows some of the light to leave the cavity to produce the laser's output beam.

Light from the medium, produced by **spontaneous emission**, is reflected by the mirrors back into the medium, where it may be amplified by stimulated emission. The light may reflect from the mirrors and thus pass through the gain medium many hundreds of times before exiting the cavity. In more complex lasers, configurations with four or more mirrors forming the cavity are used. The design and **alignment** of the mirrors **with respect to** the medium is crucial to determining the exact operating wavelength and other **attributes** of the laser system.

Get to Work

Ⅰ. Matching

Directions: *Match the words or expressions in the left column with the Chinese equivalents in the right column.*

1. antielectron   A. 声光腔
2. basic frequency   B. 二极管抽运
3. fiber sensor   C. 反电子
4. optical filter   D. 准直光束
5. photo-electron   E. 交流电
6. diode pumping   F. 光电子
7. input amplifier   G. 主频率
8. collimated beam   H. 光纤感应器
9. acousto-optic cavity I. 输入放大器
10. alternating current J. 滤光器

Ⅱ. Understanding Checking

Directions: *Choose the best answer for each statement below according to the text.*

1. A laser is made up of the following parts: a pump source, a gain medium, and _____.
   A. the pump   B. an energy source   C. laser medium   D. an optical resonator

2. Pump sources include arc lamps, light from another laser, electrical discharges, and even _____.
   A. xenon   B. explosive devices   C. a coupler   D. a high reflector

3. The gain medium is _____ by the pump source to produce a population inversion.
   A. stimulated   B. motivated   C. encouraged   D. inspired

4. Different gain media contain liquids, gases, _____, and semiconductors.
   A. dopants   B. crystal   C. solids   D. coating

5. Optical coatings _____ the reflective properties of the mirrors in the optical resonator.
   A. depend on   B. determine   C. allow for   D. take advantage of

Ⅲ. Passage Completion

Directions: *Fill in each of the blanks with one of the words or expressions in the box, making changes if necessary.*

| measurement | pulse | generate | prove | considerable |
| resonator | set-up | lamp | principle | operate |

On May 16th, 1960, the laser was born. Laser light was __1__ for the first time. The experiment __2__ by Theodore H. Maiman was astonishingly simple with a short, thick ruby rod as an active medium. The end surfaces of the rod were used as the __3__. A coiled flash 4 was used for excitation. The first laser developed by Maiman was a 5 solid-state laser.

Maiman's ruby laser could only 6 with minimal pulse repetitions. But the pulse power and energy were already quite 7 : a pulse penetrated through a packet of razor blades. This was no mere game. The number of penetrated razor blades was the relative unit of measurement for the pulse energy because there were no 8 devices at the time.

Its successors 9 themselves particularly in production technology. Tens of thousands of them do their work around the world on a higher technical level but still based on the same 10 .

Ⅳ. Translation

Directions: Complete the sentences by translating into English according to the Chinese given in brackets.

1. Some optical devices may be placed within the optical resonator to _____ _____（产生多重效应）on the laser output.

2. Some so-called lasers rely on very high optical gain without _____ _____（光返回增益介质的反馈）.

3. Since they do not use optical feedback, these devices _____ _____（不归类为激光器）.

4. Semiconductor lasers are typically very small, and can _____ _____（通过简单的电流抽运）.

5. In 1947, Willis E. Lamb and R. C. Retherford _____ _____（首次展示了受激发射）.

Ⅴ. Sentence Remaking

Directions: Simulate the following sentence patterns according to an example provided from the text.

1. ... is/are constructed from ...

E.g. A laser is constructed from three principal parts.

_____ is/are constructed from _____.

2. This determines how ...

E.g. This determines how the energy is transmitted to the medium.

This determines how _____.

3. ... is/are ... by ... to ...

E.g. The gain medium is excited by the pump source to produce a population inversion.

_____ is/are _____ by _____ to _____.

4. ... takes/take place, leading to ...

E.g. Spontaneous and stimulated emission of photons takes place, leading to the phenomenon of optical gain.

_____ takes/take place, leading to _____.

5. ... is/are crucial to ...

E. g. The design and alignment of the mirrors are crucial to determining the exact operating wavelength and other attributes of the laser system.

_____ is crucial to _____.

VI. Para. Translation

Most solid state lasers are rod lasers (laser rods), i. e., lasers with a rod-shaped doped laser crystal as the gain medium. Strictly speaking, a rod always has a cylindrical (圆柱的) shape, but the term is sometimes also used for crystals with rectangular shapes. Its ends are normally either perpendicular (垂直的) to the beam axis, or are Brewster-angled (布鲁斯特角) for suppressing (抑制) parasitic reflections and ensuring a stable linear polarization. Rod lasers can be either end-pumped or side-pumped. Cooling is the simplest for end pumping, where the outer surface can be fully surrounded by a water-cooled heat sink.

VII. Mini Imitative Writing

Directions: The following paragraph originates from the text. Please imitate it and write another one.

The pump source is the part that provides energy to the laser system. Examples of pump sources include electrical discharges, flashlamps, arc lamps, light from another laser, chemical reactions and even explosive devices. The type of pump source used principally depends on the gain medium, and this also determines how the energy is transmitted to the medium.

3.4 Internet and Lasers

Learning objectives:
In this module, you will:
- understand the ways to connect people to the Internet from the sky.
- learn about lasers' role in delivering the Internet.
- master strategies adopted by companies in broadening Internet access from the sky.

Warm-Up Activity

After reading the descriptions of the four pictures, please judge whether they are true or false.

Picture 1

Picture 2

Picture 3

Picture 4

Descriptions		True	False
Picture 1	The rate of data transmission can be increased in the optical-fiber network.		
Picture 2	Lasers can be used to make the Internet faster.		
Picture 3	Laser marketing can be furthered through online shopping malls.		
Picture 4	The laser in combination with AI is the best way for Internet development.		

Text

Facebook, among other large Internet companies, is trying to **figure out** how to connect the next billion people to the Internet.

Without access to expensive **wired** Internet connections, these people largely won't be getting online from computers, but from mobile devices. The race is on to figure out how to connect those devices from the sky—be it by satellite, weather balloon, or **drone**.

On Tuesday, Facebook CEO Mark Zuckerberg said he **envisions**

figure out 解决,算出,想出

wired *adj.* 有线的

drone *n.* 无人机

envision *vt.* 设想

a world where we can digitally send our emotions to others. But that technology is still a few years out, so it looks like now Zuck will have to **settle for** drones that shoot the Internet through lasers from the sky.

Zuckerberg gave us a **sneak** peek at some of the lasers the Facebook **Connectivity** Team are building in their project to expand Internet connectivity around the world.

settle for 勉强接受

sneak *adj.* 暗中进行的
connectivity *n.* 连接,联结

Facebook Aquila Plane

Facebook's Connectivity Lab is working on a "laser communication system" that will be able to send data to entire communities. In March, Zuckerberg confirmed that they had completed their first test flight of their drone, which will reportedly have a larger **wingspan** than a Boeing 737 (102 or 138 feet, depending on the model), and weigh less than a car. No photos or videos were released of that drone, but this newest update is the first real look at some of the technology involved in the project.

wingspan *n.* 翼展,翼幅

The project could change how the Internet is **delivered** to those who don't have access to a fixed connection. Combine this laser system with the drone technology the social network is working on, and a picture of a wireless network of Internet-connected drones flying overhead starts to **emerge**. The lasers, Zuckerberg noted, will not be **visible** when in use.

deliver *vt.* 递送,释放

emerge *vi.* 浮现,暴露
visible *adj.* 明显的,看得见的

Laser as a form of communication, which **falls under** the **umbrella** of free-space optical communication, is still a relatively new field. NASA reported using a laser to beam data to the moon at 622 megabits per second in 2013, and it's seen some interest from defense **contractors** as well.

fall under 被归入,受到(影响等)
umbrella *n.* 统称,总称
contractor *n.* 承包人,承包商

An employee at Facebook's Connectivity Lab testing the data-delivery laser

Zuckerberg and his team aren't the only ones who want to **broaden** Internet access from the sky—Google's Project Loon looks to use high-**altitude** balloons that emit Internet signals, and even Tony Stark **wannabe** Elon Musk wants to send Internet-beaming satellites to orbit Earth. And not be **outdone**, Virgin Galactic, led by Sir Richard Branson, announced a large step to sending its own Internet satellites to space, **in a partnership with** satellite company OneWeb.

Facebook hasn't given a **timeline** to when the project will be widely available, but Zuckerberg claims that when they are **put into action**, they'll be able to serve 10 percent of the world's population that is now without Internet **infrastructure**.

Keep dreaming big, Zuck.

broaden *vi. &vt.* 扩大,变阔
altitude *n.* 纬度,高度
wannabe *n.* (名人的)崇拜模仿者
outdo *vt.* 超过,胜过
in a partnership with 与……合作
timeline *n.* 时间轴,时间线
put into action 把……付诸实施,使开始工作
infrastructure *n.* 基础设施,公共建设

Get to Work

Ⅰ. Matching

Directions: *Match the words or expressions in the left column with the Chinese equivalents in the right column.*

1. periodic focusing   A. 角色散
2. optical lock-on   B. 闭合共振器
3. argon flash   C. 空对空激光测距
4. excited electron   D. 绝对误差

5. flexible mirror	E. 偏差补偿
6. deviation compensation	F. 光学锁定
7. absolute error	G. 活动反射镜
8. angle dispersion	H. 周期聚焦
9. air-to-air laser ranging	I. 受激电子
10. closed resonator	J. 氙气闪光灯

Ⅱ. Understanding Checking

Directions: *Fill in the blanks with no more than three words according to the text.*

1. It is possible to connect people to the Internet by satellite, weather balloon, or _____.

2. The lasers to expand Internet connectivity around the world are being built by the _____.

3. Facebook's drone will weigh less than _____.

4. The lasers used by Facebook will be _____ when in use.

5. Elon Musk plans to send Internet-beaming _____ to orbit Earth.

Ⅲ. Passage Completion

Directions: *Fill in each of the blanks with one of the words or expressions in the box, making changes if necessary.*

| as | weld | so that | replace |
| shorten | operation | option | save |

Scanner welding is a very fast and dynamic joining technology that shows its strengths in more than just car making. In general, it is an interesting __1__ for quantities greater than 100,000 units per year or when many welding __2__ are acquired for each component. The time saved by moving faster from one welding spot to the other can __3__ the processing time by up to three quarters. A scanner welding station can therefore __4__ several conventional welding stations, which not only __5__ space but also, as a result of the laser's basic flexibility, allows various components to __6__ in one station.

Scanner welding illuminates nonproductive downtime caused by positioning the welding tool. This advantage becomes more significant __7__ the distances between the welding spots become greater and the number of weld spots per component increases. There is a positive side effect: without losing any time, the seam sequence can be selected individually __8__ the heat input is evenly distributed.

Ⅳ. Phrase Translation

Directions: *Complete the sentences by translating into English according to the Chinese given in brackets.*

1. This causes rapid heating and _____

（周围空气的爆炸性扩张）.

2. The device makes noise similar to ＿＿＿＿＿＿＿＿＿＿
（伴随闪电的雷声）.

3. When ＿＿＿＿＿＿＿＿＿＿
（这种现象发生在某些科学实验中）, it is referred to as a "plasma mirror"（等离子镜）.

4. A laser beam would not ＿＿＿＿＿＿＿＿＿＿
（肉眼所不能见）in the near vacuum of space.

5. In 1953, Charles H. Townes and others produced the first microwave amplifier, a device that ＿＿＿＿＿＿＿＿＿＿
＿＿＿＿＿＿＿＿＿＿＿＿＿＿＿＿＿＿＿＿＿＿（工作原理与激光类似）.

Ⅴ. Sentence Remaking

Directions: *Simulate the following sentence patterns according to an example provided from the text.*

1. ... especially in ... , is/are ...

E. g. The representation of lasers in popular culture, especially in science fiction and action movies, is often misleading.

＿＿＿＿＿＿, especially in ＿＿＿＿＿, is/are ＿＿＿＿＿＿＿＿＿＿.

2. It is ... to ... than ...

E. g. It is far easier and cheaper to build infrared laser diodes than visible light laser diodes.

It is ＿＿＿＿＿ to ＿＿＿＿＿＿＿＿＿ than ＿＿＿＿＿＿＿＿.

3. ... to ... is likely to ...

E. g. Putting enough dust in the air to make the beam visible is likely to be enough to "break" the beam and trigger the alarm.

＿＿＿＿＿＿ to ＿＿＿＿＿＿ is likely to ＿＿＿＿＿＿＿.

4. ... , while ...

E. g. *Goldfinger*'s laser makes a whirring electronic sound, while a real laser produces a fairly heat-free and silent cut.

＿＿＿＿＿＿＿＿＿＿＿, while ＿＿＿＿＿＿＿＿＿＿＿.

5. ... is ... as it is/ they are ...

E. g. Laser light is not perfectly parallel as it is sometimes claimed.

＿＿＿＿ is ＿＿＿＿＿＿＿＿＿ as it is/ they are ＿＿＿＿＿.

Ⅵ. Para. Translation

Earlier this month, NASA successfully tested a new technology that allows the agency to beam images, videos, and other data from space to Earth on lasers rather than the old fashioned way—on radio waves. The benefit of the laser technology is greater transmission speed. The increase could be anywhere from 10 to 1,000 times as fast. "NASA missions collect an enormous amount of data out in space," said Matt Abrahamson, the project manager working for NASA. "Laser communications is a faster alternative for getting those data to the ground."

Ⅶ. Mini Imitative Writing

Directions: The following paragraph originates from the text. Please imitate it and write another one.

Facebook hasn't given a timeline to when the project will be widely available, but Zuckerberg claims that when they are put into action, they'll be able to serve 10 percent of the world's population that is now without Internet infrastructure.

CAREER SPEAKING

Introducing Enterprises

Ⅰ. Warm-Up

Model

(A: manager　　B: guest)

B: Would you like to let me know how many departments you have?

A: Yes, we have 5—Planning Department, Sales Department, Production Department, Personnel Department and **Accounting Department**.

B: I know. You must have a lot of employees.

A: Yes. There are 2,700, including both administrative staff and technical staff. 40% of them have a Master's degree.

B: As I know, your employees have good **labor protection** and a favorable salary.

A: Right. In addition, we have **recreational facilities**, a medical center and **dormitories** for the single employees.

B: How about their promotion?

A: Every one has an opportunity to be promoted **on the basis of** their performance and **merit**.

B: Your employees are so lucky to work here.

A: All members of the company are part of a large family. Therefore, it is important for us to preserve **harmony** and teamwork.

B: Your words are really impressive!

Notes

Accounting Department	财务部
labor protection	劳动保护
recreational facilities	娱乐设施
dormitory	n. 宿舍
on the basis of	在……的基础上
merit	n. 功绩
harmony	n. 和睦，和谐

Ⅱ. Matching

Directions: Please choose the corresponding sentences from the box to fill in the blanks, then play roles with your partner.

(A: manager   B: guest)

A: Welcome to our company. Please be seated.

B: Thank you. Your company is very reputed in this city. _____ Well, when did it come into being?

A: It was established in 1990. _____

B: Oh, it is still a young company, but your development of optical fiber communications has been remarkable.

A: Indeed. _____ And the demand is getting greater and greater.

B: _____

A: You're right. _____

B: Your company really has strength.

> S1: We have been in the trade for more than 20 years.
> S2: As far as technology is concerned, I believe our company is among the best in the world.
> S3: I can see you are really proud of your company.
> S4: I heard much praise to your company.
> S5: Our products are well-accepted in the world market.

Ⅲ. Closed Conversation

Directions: Please fill in the blanks according to the Chinese, then play roles with your partner.

(A: secretary   B: guest)

B: This is my first visit to your company. I'm lucky to have the chance to do so.

A: Welcome. _____

(我现在就带您四处看看。参观后您会对我们的公司有更深的了解。)

  B：I'm sure I'll.

  A：This way please. _____

(我们从车间开始吧。)

  B：_____(生产线是全自动的吗？)

  A：Yes. _____(几乎每一道工艺都是由电脑控制的。)

  B：It saves quite a lot of manpower.

  A：Indeed. _____

(通过自动化我们的效率增加了30%，成本降低了15%。)

  B：_____

(现在您能给我展示一下如何操作这台机器吗？)

  A：No problem. _____

(我需要做的就是按下这儿的电钮。您可以试试。)

  B：Very convenient. _____(产品的销路如何？)

  A：_____

(我们的产品近几年销售得很好，而且销量一直在稳步地增长。)

  B：_____(你们有产品目录和价格表吗？)

  A：Yes, after the visit I'll give them to you.

  B：Thank you. _____(那栋白色的5层楼房是干什么的？)

  A：_____(在那儿我们有娱乐室、医疗中心和单身宿舍。)

  B：Wow! Terrific!

  A：Now let's go to the R&D Department.

Ⅳ. Semi-Open Conversation

  Directions: *Please fill in the blanks with your own words, then play roles with your partner.*
  (A：secretary B：guest)

  A：Welcome to our company. _____. Please ask questions at any time.

  B：OK.

  A：_____.

  B：Oh, it is a large company with a long history. _____

  A：Yes, we have 7-_____

  B：Too many departments! _____

  A：We manufacture household appliances.

  B：Well, what is your market share?

  A：_____

  B：It sounds good.

Ⅴ. Live Show

Challenge 1

Directions: Lucy Green, a businesswoman of an American firm, is visiting a Chinese company that makes and sells optical instruments in case there is possibility to establish a long-term cooperation. Wang Fang, is in charge of showing her around. Please act it out with your partner or partners. The model and the useful expressions below are just for your reference.

Challenge 2

Directions: The students in groups discuss how to give a general picture of a company including its history, size, employment, management, business scope, product quality, annual output, production process and other aspects. One student of each group makes an oral report to the whole class. The model and the useful expressions below are just for your reference.

Model

(A: secretary   B: Mr. Smith)

A: Welcome, Mr. Smith. I'll show you around. I am sure you will know our company better after the visit.

B: That'll be most helpful.

A: If you have any questions, please feel free to ask. I will answer any questions that I can, except for those concerning corporate secrets.

B: OK.

A: Our site covers an area of approximately 30,000 square meters and we have about 1,000 employees.

B: It's much larger than I expected.

A: Do you see that white building on your right?

B: Yes.

A: That's our **office block**. We have four departments there: Advertising, Purchasing, Marketing and **R&D**.

B: What sort of products do you deal with?

A: We **specialize in** manufacturing **a wide range of** telecommunication **devices** and export them all over the world.

B: Good. Where is your factory?

A: Mr. Smith, this way please. On the opposite of the office block, that gray house of three storeys is our factory. All parts are manufactured and **assembled** there.

B: It looks rather **magnificent**! Does the factory work with everything from **raw** materials to finished products?

A: Yes. We've introduced new technology and automation system, so the whole production takes about only 25 minutes from beginning to end.

B：An efficient factory! How do you ensure quality control?

A：All products have to pass strict inspection before they go out. We control quality by computer. And of course, the final inspection is done by the engineers.

B：Would you mind telling me the **annual output value** approximately?

A：The annual output value varies between 7 to 8 million Yuan RMB.

B：Terrific! Do you spend a lot on research and development each year?

A：Yes, we invest almost one third of the profits in it.

B：Would it be possible for me to have a closer look at your samples?

A：No problem. I'll take you down to our showroom. Follow me, please.

B：I'd like to have your latest **catalogues** or something that tells me about your company and your products.

A：OK. Here you are.

Notes

office block	办公大楼
R&D	研究与开发（research and development）
specialize in	专攻，专门研究
a wide range of	各式各样的
device	n. 装置，设备
assemble	v. 装配
magnificent	adj. 壮观的
raw	adj. 未加工的
annual output value	年产值
catalogue	n. 目录

Useful expressions

1. 客户咨询

How many departments do you have in your company?

What are your company's main products?

What was the total amount of your sales last year?

2. 接待来访

We are honored to show our company to such a distinguished group of guests.

I will give you a complete picture of our operation.

Maybe we could start with the Designing Department.

3. 公司概况

We now command 35% of the market in home appliances.

The company was established in 1965, and we have about 5,500 employees now.

Our company is one of the world's biggest manufacturers of optical communication receivers, with over 30 branches at home and abroad.

4. 产品说明

There is a good market for these articles.
One of the main strengths is the quality of our products.
All products have to pass strict inspection before they go out.

5. 生产过程

We've increased our efficiency by 20% through automation.
Production takes about 25 minutes from beginning to end.
What the workers should do is to push the button here.

WRITING

Letter of Congratulation

在西方，当得知亲友或同事有订婚、升迁、获奖等喜庆之事时，往往会写信表示祝贺。内容通常包含三个方面：一，直接点明事件，表达喜悦之情；二，对事件进行积极评价，赞扬、祝福对方；三，表示真诚的祝贺与期望。此外，凡是遇到重大的喜庆节日，如圣诞节、新年等，人们也会写祝贺信，但语言会更简洁，寥寥数语即可。祝贺信传达的是美好温馨的情感，因此语言应热情洋溢，亲切自然。

Sample 1

Dear Sue,

We are very glad to hear the news of your promotion to Sales Manager of the company. Congratulations!

It is a well-deserved recognition(认可) of your hard work and great achievements. We're encouraged by your efforts rewarded and we are proud of you.

We send you our best wishes for success and happiness in your future.

Yours cordially,
Helen

Sample 2

Dear Mary,

I offer you my warmest congratulations on your **engagement**. That's great news about you and your **fiancé** Bob! I know you have common interests and tastes in every way. I send you both every good wish.

May the years ahead bring you joy and **contentment**!

Yours,
Wang Xiao

Notes

engagement	n. 订婚
fiancé	n. [法]未婚夫
contentment	n. 满足

Useful expressions

1. I am so pleased to hear that …
2. I offer you my warmest congratulations on your…
3. Heartfelt congratulations on…
4. Accept our heartiest congratulations and all good wishes.
5. Please accept our sincerest congratulations and very best wishes for all the good future.

Challenge

Please write a letter of congratulation to your colleague Susan, who has been rewarded due to her outstanding achievements.

EXTENSIVE READING

3.5　A Prospectus about Laser Machines

TruLaser Cell Series 3000

(Flexible, complete solutions for small components)

The TruLaser Cell Series 3000 offers a flexible solution for 2D and 3D processing of small parts. Typical applications are to be found in the field of medical technology, in precision mechanics and in the electronic industry — wherever the emphasis is on high precision, quality and safety.

LASERCELL 1005

(The entire world of flexible laser material processing)

Whether you have a job shop with constantly changing tasks, or need to meet strict requirements as in the aerospace industry or serial production in the automotive industry: the LASERCELL 1005 unites flexibility and productivity — also in the rough industrial environment.

LASERCELL 1000

(Indispensable for round parts)

Radial seams, axial seams and everything else in between — the LASERCELL 1000 is geared for these applications. On this compact machine, the laser, the electrical parts and fixtures share a common base plate. An extensive range of accessory equipment is available, as well as clearly defined interfaces to all conventional quality assurance systems.

LASERCELL 6005

(Flexibility, also for extra large parts)

The LASERCELL 6005 offers just the solution for the laser processing of work pieces as big as automobiles. Whether you are laser cutting or laser welding: the system enables cost-efficient operation coupled with full flexibility in terms of laser power and automation components.

TruLaser Robot 5020

(A module for complex work pieces)

The TruLaser Robot 5020 is a complete robot cell for 3D welding of complex work pieces and mass production. A flexible laser light cable guides the laser beam to the processing optics at the robot arm.

Exercise

Which can do what?

Directions: Please link the laser machine on the left with its corresponding functions and features on the right by drawing a line according to the text.

Laser Machines
TruLaser Cell Series 3000
LASERCELL 1005
LASERCELL 1000
LASERCELL 6005
TruLaser Robot 5020

Functions and Features
complex work pieces
axial seams made
coping with extra large parts
3D processing of small parts
a shared base plate
cost-efficient operation
help of the robot arm
working in rough industrial environment
applications in precision mechanics
serial production in the auto industry

3.6 Using Lasers

The use of lasers in medicine is complex, and your facial plastic surgeon is trained in the use of lasers and understands how and when to use a laser. Your surgeon will decide if a laser is appropriate, and which laser is best for the situation.

In medicine, physicians can use lasers to make incisions, vaporize tumors, close blood vessels, or even treat skin **wrinkles**. The laser makes it possible to change tissue without making an incision. So a surgeon can treat birthmarks or damaged blood vessels, remove port-wine stains, and shrink facial "spider veins" without major surgery.

Is it any wonder that many facial plastic surgeons use lasers on a routine basis? They use the laser as a "light scalpel". The tissue is left sterile, and bleeding is greatly reduced. When the laser is used to treat port-wine stains, no cuts are made. The laser energy penetrates through the skin to shrink the abnormal blood vessels that are the cause of these marks.

There are different types of laser surgery, e.g. laser skin peeling. Lasers can be used to reduce wrinkles around the lips or eyes, even the entire face, softening fine wrinkles and removing certain blemishes on the face.

Another example is laser removal of birthmarks and skin lesions. Port-wine stain birthmarks respond remarkably well to laser treatment. The abnormal blood vessels that cause these marks are reduced in size by the laser. This results in a lightening of the treated area. Skin growths, facial "spider veins", and some tattoos respond to laser surgery. Most situations take more than one laser treatment, but some respond to a single treatment.

The facial plastic surgeon often uses the minimum laser intensity. The low intensity requires many treatments. However, the low intensity also preserves as much of the healthy tissue as possible. This produces an aesthetically pleasing result. Many of these laser surgeries are performed as outpatient treatments in hospitals or offices.

After your surgeon has indicated that a laser can be helpful in the surgery, your surgeon will explain the laser of choice and what can be accomplished. As with all surgery, the laser has its limitations. Often the results are spectacular. Your surgeon will give you the best judgment for the particular procedure.

Some surgeons may use local anesthetics to numb the treated area before the surgery. Surgery can sometimes be done in the surgeon's office; other times the surgeries are performed in outpatient facilities at a clinic or hospital. Your surgeon will decide on the appropriate method, dictated by the nature of the surgery.

Because safety is a major component of laser use, your surgeon will describe safety precautions before the surgery. If you are treated with a local anesthetic, you will be required to wear protective glasses or goggles during laser use.

Exercise

Directions: Choose the best answer from the four choices.

1. What is the meaning of the word "wrinkle" in Para. 2?
   A. Line.   B. Scar.   C. Color.   D. Shortcoming.
2. How are port-wine stains treated through a laser?
   A. A laser makes an incision.
   B. Laser skin peeling is a good solution.
   C. Some cuts are made by a laser.
   D. The laser passes into the skin to treat the abnormal blood vessels.
3. Which of the following statements is false according to the passage?
   A. Laser skin peeling can remove certain flaws on the face.
   B. The laser works in reducing the marks that cause the abnormal blood vessels.
   C. The laser plays a part in treating some tattoos.
   D. A single laser treatment is enough in most cases.
4. Why is low laser intensity often used in facial plastic surgery?
   A. Low laser intensity is usually demanded by patients.
   B. The surgeon wants to make the operation cost-effective.
   C. Low laser intensity can lead to be healthier and better-looking.
   D. Its effectiveness is a determining factor.
5. Which one is true about local anesthetics during the laser use?
   A. Local anesthetics are necessary in most operations.
   B. With a local anesthetic, the treated area is numbed.
   C. Some risks about local anesthetics are not included in the surgeon's precaution before the surgery.
   D. Protective glasses or goggles have nothing to do with a local anesthetic.

TRANSLATION

Translation of Noun Clauses

名词性从句的翻译

英语的名词性从句包括主语从句、宾语从句、表语从句和同位语从句四种。

1. 主语从句的译法

一般采用顺译法和倒译法。

顺译法：按照句子的排列顺序，先译主句，再译主语从句。

It is obvious **that A. C. motors are superior to D. C. motors in many applications.**

显然，交流电机在许多应用场合比直流电机优越。

倒译法：常用于有形式主语 it 的结构中，先译真正的主语从句，再译谓语。

It is a fundamental rule of physics **that energy can neither be created nor destroyed.**

能量既不能被创造也不能被消除，这是物理学的一个基本原则。

2. 宾语从句的译法

往往采用顺译法，包括以下两种情况。

1) 动词后的宾语从句

It's necessary to know **that voltage is used up in components, but not in wires.**

有必要知道，电压消耗在元件上，而不是导线上。

2) 介词后的宾语从句

The scientific research depends on **whether there are proper subjects.**

科学研究取决于是否有合适的实验对象。

3. 表语从句的译法

一般可以按原文顺序翻译，多采用"是"的句式。

This is **where the basic principle lies.**

这是基本原理所在。

4. 同位语从句的译法

1) 译成定语

Obviously there was little probability **that this experiment would succeed.**

很显然，实验成功的可能性极小。

2) 省略连接词前的名词

The advantage of this device is due to the fact **that it has an automatic temperature control.**

这台设备的优点在于它是自动调温的。

（the fact 省去不译）

3) 译成独立的句子

We recognize the fact **that the operational amplifier is the most useful single device in an-**

alog electronic circuitry.

我们认同这个事实,运算放大器是模拟电子电路中最有用的独立器件。

Challenge

Translate the following sentences into Chinese.

1. An important consequence of this is that thermal lensing effects inevitably become strong for high output power levels.

2. This is due to the fact that fewer mirrors are required to deliver the laser beam to the cutting head.

3. It is evident that current cannot flow without voltage.

4. Application-based manufacturing differences mean that laser diode and LED products are constructed in fundamentally different ways.

5. It's a good idea that various broadband communication services can be provided.

TASK

Laser Construction

Target	Analyze and describe laser construction.				
Requirements	Make an English presentation according to lesson 3. 4. Produce a micro-video to express professional knowledge. Add accurate Chinese and English subtitles to the video.				
Contents	Laser Construction: Pump source Gain medium Optical resonator				
Group: _____	Project	Name	Software	Time	Score
	Material collection				
	Script				
	PPT				
	Reader				
	Subtitle				
	MP4				
Operating environment	XP/Win7/Win8/ Mobile phone				
Product features					

Script reference

标题	专业内容(讲稿与中、英文字幕)	PPT 中需显示的文字	表现形式	素 材
每个知识点的要点或者小节名字	做报告时的英文讲稿 视频中的中、英文字幕	需要特别提醒的文字	讲解到某个地方或某句话时,出现的图片、动画或者文字	图片或者视频
Contents	Laser is constructed from three principal parts: An energy source (usually referred to as the pump or pump source), a gain medium or laser medium, and two or more mirrors that form an optical resonator. 激光由三个主要部分组成:能源(通常被称为泵或泵源),增益介质或激光介质,两个或两个以上的镜子组成的光学谐振腔。	Laser Construction: Pump source Gain medium Optical resonator	主要内容和激光器结构图	图 1 激光器的结构
1. Pump source 泵源	The pump source is the part that provides energy to the laser system. 泵源是向激光系统提供能量的部分。		强调图中泵源部分	图 1 激光器的结构
	Examples of pump sources include electrical discharges, flash lamps, arc lamps, light from another laser, chemical reactions and even explosive devices. 泵源包括放电、闪光灯、弧光灯、另一激光器的光束、化学反应,甚至爆炸装置等。	Electrical discharges, flash lamps, arc lamps, light from another laser, chemical reactions and even explosive devices.	依次出对应的图	图 2 电子放电 图 3 闪光灯

续表

标题	专业内容(讲稿与中、英文字幕)	PPT 中需显示的文字	表现形式	素 材
				图 4 弧光灯
				图 5 另一激光器的光束
				图 6 光化学反应
				图 7 爆炸
2. Gain medium 增益介质	The gain medium is the major determining factor of the wavelength of operation, and other properties of the laser. 增益介质是激光工作波长及其他特性的决定性因素。	Gain medium	强调图中增益介质部分	图 1
	The gain medium is excited by the pump source to produce a population inversion, and it is in the gain medium that spontaneous and stimulated emission of photons takes place, leading to the phenomenon of optical gain, or amplification. 泵源激活增益介质,产生粒子数反转,在增益介质中,会发生光子的自发和受激发射,导致光学增益现象,即放大。		强调图中增益介质部分	图 1

续表

标题	专业内容（讲稿与中、英文字幕）	PPT 中需显示的文字	表现形式	素　材
	Examples of different gain media include： 增益介质包括：			
	Liquids，such as dye lasers. The exact chemical configuration of the dye molecules determines the operation wavelength of the dye laser. 液体，如染料激光器。染料分子的精确化学构造决定染料激光的工作波长。	Liquids，such as dye lasers.	液体图	图8 液体
	Gases，such as carbon dioxide and mixtures such as helium-neon. These lasers are often pumped by electrical discharge. 气体，如二氧化碳和氦氖混合物。这类激光器通常通过放电抽运。	Gases，such as carbon dioxide and mixtures such as helium-neon.	气体图	图9 气体
	Solids，such as crystals and glasses. The solid host materials are usually doped with an impurity. 如晶体和玻璃等固体。固体基质材料通常掺杂杂质。	Solids，such as crystals and glasses.	晶体和玻璃	图10 晶体
	Semiconductors，a type of solid，in which the movement of electrons between material with differing dopant levels can cause laser action. 半导体是固体的一种，电子在不同掺杂水平的材料间流动能产生激光作用。	Semiconductors，a type of solid.	半导体	图11 半导体
3. Optical resonator 光学谐振器	The optical resonator，or optical cavity，in its simplest form is two parallel mirrors placed around the gain medium which provide feedback of the light. 光学谐振器或光学腔的最简单形式是在增益介质周围平行放置着两个反射镜，反馈光束。	The optical resonator，or optical cavity，in its simplest form is two parallel mirrors placed around the gain medium which provide feedback of the light.	强调图中光学谐振腔	图1

续表

标题	专业内容（讲稿与中、英文字幕）	PPT中需显示的文字	表现形式	素材
	The mirrors are given optical coatings which determine their reflective properties. Typically one will be a high reflector, and the other will be a partial reflector. The latter is called the output coupler, because it allows some of the light to leave the cavity to produce the laser's output beam. 反射镜的镀膜决定其反光性能。通常一面是全反射镜，另一面是部分反射镜。后者被称为输出耦合器，因为它允许部分光离开腔，输出激光光束。	Typically one will be a high reflector, and the other will be a partial reflector.	强调两块镜子	图1
	Light from the medium, produced by spontaneous emission, is reflected by the mirrors back into the medium, where it may be amplified by stimulated emission. The light may reflect from the mirrors and thus pass through the gain medium many hundreds of times before exiting the cavity. 由自发辐射产生的介质光经镜子反射重回到介质，经受激发射后可以被放大。光能被镜子反射，因此在溢出腔外前数百次地穿越增益介质。	Light from the medium, produced by spontaneous emission, is reflected by the mirrors back into the medium, where it may be amplified by stimulated emission. The light may reflect from the mirrors and thus pass through the gain medium many hundreds of times before exiting the cavity.	强调光在溢出腔外前数百次地穿越增益介质	图12 光路
	In more complex lasers, configurations with four or more mirrors forming the cavity are used. The design and alignment of the mirrors with respect to the medium is crucial to determining the exact operating wavelength and other attributes of the laser system. 在更为精密的激光器中，谐振腔由4个或更多的反射镜构成。与介质相关的反射镜的设计和排列决定激光系统的精确工作波长及其他特性。	In more complex lasers, configurations with four or more mirrors forming the cavity are used.	谐振腔由4个或更多的反射镜构成	图13 多个反射镜

UNIQUE ENGLISH

Funny Chinglish

Directions: There is something wrong in the English signs and notices of the following pictures. Have a try to correct them.

Picture 1

Picture 2

Picture 3

Picture 1: Fire fight safety device. Please don't move at will.（消防安全设施 请勿随意搬运）

Picture 2: No entrance.（顾客止步）

Picture 3: Cherishing environment. Everyone's duty.（爱护环境 人人有责）

Unit 4

Optical Detection

In this unit, you are going to learn the following contents:

READING IN DEPTH

4.1　Photoelectric Effect

4.2　Photodiode

4.3　Optical Detection Techniques

4.4　CCD

CAREER SPEAKING

Making Calls

WRITING

Letter of Apology

EXTENSIVE READING

4.5　LED Technology

4.6　Optical Storage Technology

TRANSLATION

Translation of Attributive Clauses

TASK

Optical Detection

UNIQUE ENGLISH

Exotic Sentences

READING IN DEPTH

4.1 Photoelectric Effect

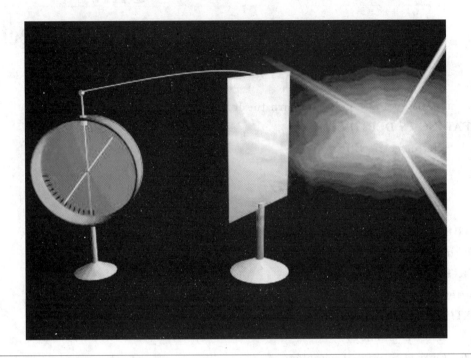

Learning objectives:

In this section, you will:
- understand the definition of the photoelectric effect.
- try to explain the photoelectric effect by the wave or quantum theory.
- comprehend the experiment: no matter how bright a red light you have, it will not produce a current in a metal, but even a very dim blue light will result in a current flowing.
- A: These results can't be explained if light is thought of as a wave, WHY?
- B: With light as photons, these results can be explained. Try to explain.
- understand that a good way to think of the photoelectric effect is like a full car park with lots of really bad drivers.

Warm-Up Activity

Do you know the names of the applications based on the photoelectric effect?

Unit 4 Optical Detection

Picture 1

Picture 2

Picture 3

Picture 4

Picture 5

Text

In the 20th century physics, two ideas stand out as being totally revolutionary: **relativity** and **quantum theory**. Although Einstein is best-known for his theory of relativity, he also played a major role in developing quantum theory. And it was his contribution to quantum theory — explaining the **photoelectric effect** — which won Einstein his **Nobel Prize** in 1921.

The photoelectric effect is the name given to the **observation** that when light is shone onto a piece of **metal**, a small current flows through the metal. The light is giving its energy to electrons in the **atoms** of the metal and allowing them to move around, producing the current. However, not all colors of light affect metals in this way. No matter how bright a red light you have, it will not produce a current in a metal, but even a very **dim** blue light will result in a current flowing. The problem was that these results can't be explained if light is thought of as a wave. Waves can have any amount of energy you want—big waves have a lot of energy, small waves have very little. And if light is a wave, then the **brightness** of the light affects the amount of energy — the brighter the light, the bigger the wave, the more energy it has. The different colors of light are defined by the amount of energy they have. If all else is equal, blue light has more energy than red light with yellow light somewhere in between. But this means that if light is a wave, a dim blue light would have the same amount of energy as a very bright red light. And if this is the case, then why won't a bright red light produce a current in a piece of metal as well as a dim blue light? Einstein realized that the only way to explain the photoelectric effect was to say that **instead of** being a wave, as was generally accepted, light was actually **made up of** lots of small packets of energy called photons that behaved like particles. Einstein wasn't the first person to use the idea of photons, but he was the first to make it the starting point of an **explanation**.

With light as photons, Einstein showed that red light can't **dislodge** electrons because its individual photons don't have enough energy — the impacts are just not large enough to shift the electrons. However, blue light can dislodge electrons — each **individual** photon has more energy than the red photon. And photons of **ultraviolet light**, which have yet more energy, will give electrons enough energy

---

relativity *n.* 相对性，相关性，[物]相对论
quantum theory 量子论
photoelectric effect 光电效应
Nobel Prize 诺贝尔奖
observation *n.* 观察，观测，观察资料（或报告）
metal *n.* 金属
atom *n.* 原子

dim *adj.* 暗淡的，模糊的，无光泽的

brightness *n.* 光亮，明亮，聪明

instead of 代替，而不是
be made up of 由……组成
explanation *n.* 解释，解说，说明
dislodge *v.* 驱逐

individual *adj.* 个别的，单独的，个人的
ultraviolet light 紫外线

to **whizz** away from the metal altogether. A good way to think of the photoelectric effect is like a full car park with lots of really bad drivers. There is a car parked in a space, and lots of other drivers want that space. To get it they can try knocking the parked car out of the way, but they can only manage to hit it one car at a time. A tiny red mini just won't have the energy to knock the parked car out of the parking space, but a big blue van will. And imagine hitting the parked car with a big ultraviolet **lorry**— the parked car is most likely going to move far enough to **collide** with something else. Returning to light and electrons, there is never really just one photon of light at a time. A bright light **emits** lots of photons, but it doesn't matter how bright a red light gets; red photons will still not be able to **budge** a single electron. This is like having a car park full of red minis each **randomly** hitting a parked car in turn — there will be a lot of **dents** but the parked car will remain where it is. However, even a dim blue light will **shift** some electrons — we know that even one blue van will be able to move the parked car.

whizz v. (使)飕飕作声

lorry n. 卡车, 货车
collide vi. 碰撞, 抵触

emit vt. 发出, 放射
budge v. 移动

randomly adv. 胡乱地
dent n. 凹, 凹痕
shift vt. 移动

Einstein's explanation of the photoelectric effect was just the start of an **avalanche** of discoveries that became quantum theory. In this theory, light is not just a particle and not just a wave: it can be one or the other, depending on how it is measured. And it was discovered later that even the electrons are not just particles but are waves too.

avalanche n. 雪崩

Get to Work

I. Matching

Directions: Match the words or expressions in the left column with the Chinese equivalents in the right column.

1. relativity   A. 频率与周期
2. quantum theory   B. 紫外光
3. ultraviolet light   C. 电磁波
4. electromagnetic wave D. 量子理论
5. frequency and period E. 相对论

II. Understanding Checking

Directions: Mark Y (for YES) if the statement agrees with the information given in the text; N (for NO) if the statement contradicts the information given in the text; NG (for NOT GIVEN) if the information is not given in the text.

(   )1. The quantum theory is the name given to the observation that when light is shone onto a piece of metal, a small current flows through the metal.

(   )2. Although Einstein is best-known for his theory of relativity, he also played a major role in developing quantum theory.

(   )3. No matter how bright a blue light you have, it will not produce a current in a metal, but even a very dim red light will result in a current flowing.

(   )4. However, blue light can dislodge electrons — each individual photon has more energy than the red photon.

(   )5. It was discovered later that even the electrons are not just particles but are waves too.

Ⅲ. Passage Completion

Directions: Fill in each of the blanks with one of the words or expressions in the box, making changes if necessary.

finally	begin	more	red	turn
observe	once	measure	through	respect

The experiment itself is now performed in appropriate physics classes at most colleges and universities, and possibly in some high schools. It isn't difficult to perform, and if proper care is used, will give quite accurate results.

We __1__ with a photoelectric tube (or phototube), which is a vacuum tube containing a metal plate curved into a half-cylinder (the anode) and a thin wire electrode (the cathode) along the axis of the cylinder. We use a very sensitive meter, called a galvanometer (G), to __2__ the current passing through the tube, and a variable voltage source (V) to control the voltage applied between the two electrodes inside the tube. __3__, we add a light source (usually a mercury vapor lamp, but it can be an arc light or even a tungsten filament light bulb), and a colored filter to limit the light striking the photoelectric tube to a single frequency.

With no light, of course, the current through the tube will be zero, regardless of the applied voltage. (Yes, there is a slight capacitance between the metal plate and the wire, and a high enough voltage will cause an arc. But the capacitance is very slight and will charge quickly, and we won't be using anywhere near enough voltage to cause any problems.) So we __4__ on the light, place one of our colored filters between the light and the phototube, and begin our measurements.

The first thing we note is that if we reverse the polarity of the voltage source V, increasing the voltage or the intensity of the light will increase the current flow. This seems logical; the light is "kicking" electrons off the metal plate, and they are attracted to the now positively-charged wire electrode. The __5__ light, the more electrons and hence the higher the current.

However, when we make the wire electrode negative with __6__ to the metal plate as shown in this figure, we begin to __7__ some interesting effects. First, __8__ light, infrared (IR) radiation and anything of a lower frequency will not excite the phototube enough to enable current to flow. You pretty much have to get up at least to green light before you can measure a reaction. (This may vary depending on the specific metal coating the surface of the curved electrode.)

__9__ you find a filter whose color allows current to flow __10__ the tube, you can begin to take your measurements.

Ⅳ. **Translation**

Directions: *Complete the sentences by translating into English according to the Chinese given in brackets.*

1. In 1887, the German physicist Heinrich Rudolf Hertz discovered an _____ _____ (物质的有趣的性质).

2. The main relationship developed in this chapter was the one between the incident optic power and _____ (光电检测器中产生的电流).

3. With no light, of course, the current through the tube will be zero, _____ _____ (不考虑外加电压).

4. The different colors of light _____ (根据……定义的) by the amount of energy they have.

5. The first thing we note is that if we _____ ____ (改变电压源的极性), increasing the voltage or the intensity of the light will increase the current flow.

4.2 Photodiode

> **Learning objectives:**
> In this module, you will:
> ● understand the meaning of the word "photosensor".
> ● learn about photodetector types.
> ● learn about photodetector properties.

Warm-Up Activity

Do you know the names of the applications based on the photoelectric effect?

Picture 1　　Picture 2

Picture 3　　Picture 4　　Picture 5

Text

Photosensors or **photodetectors** are sensors of light or other electromagnetic energy. A photodetector has a p-n junction that converts light photons into a current. The absorbed photons make **electron-hole pairs** in the **depletion region**. Photodiodes and photo transistors are a few examples of photo detectors. Solar cells convert some of the light energy absorbed into electrical energy.

photosensor n. 光敏元件,光传感器,光敏器件
photodetector n. 光电探测器,光检测器
electron-hole pair 电子空穴对
depletion region 耗尽区

TYPES

Photodetectors may be classified by their mechanism for detection:

Photoemission or photoelectric effect: Photons cause electrons to transition from the conduction band of a material to free electrons in a vacuum or gas.

Thermal: Photons cause electrons to transition to mid-gap states then decay back to lower bands, inducing phonon generation and thus heat.

Polarization: Photons **induce** changes in polarization states of suitable materials, which may lead to change in index of refraction or other polarization effects.

Photochemical: Photons induce a chemical change in a material.

Weak interaction effects: Photons induce secondary effects such as in photon drag detectors or gas pressure changes in Golay cells.

Photodetectors may be used in different configurations. Single sensors may detect overall light levels. A 1-D array of photodetectors, as in a spectrophotometer or a line scanner, may be used to measure the distribution of light along a line. A 2-D array of photodetectors may be used as an image sensor to form images from the pattern of light before it.

A photodetector or array is typically covered by an illumination window, sometimes having an anti-reflective coating.

Properties

There are a number of **performance metrics**, also called figures of merit, by which photodetectors are characterized and compared.

Spectral response: The response of a photodetector as a function of photon frequency.

Quantum efficiency: The number of carriers (electrons or holes) generated per photon.

Responsivity: The output current divided by total light power falling upon the photodetector.

Noise-equivalent power: The amount of light power needed to generate a signal comparable in size to the noise of the device.

Detectivity: The square root of the detector area divided by the noise equivalent power.

Gain: The output current of a photodetector divided by the current directly produced by the photons incident on the detectors, i.e., the built-in current gain.

Dark current: The current flowing through a photodetector even in the absence of light.

Response time: The time needed for a photodetector to go from 10% to 90% of final output.

Noise spectrum: The intrinsic noise voltage or current as a function of frequency. This can be represented in the form of a noise spectral density.

Nonlinearity: The RF-output is limited by the nonlinearity of the photodetector.

dark current 暗流,暗电流,无照电流(光电倍加管中)
response time 响应[反应]时间
noise spectrum 噪声谱
nonlinearity n. 非(直)线性(特性)

Get to Work

Ⅰ. Matching

Directions: Match the words or expressions in the left column with the Chinese equivalents in the right column.

1. potential barrier   A. 辐射通量
2. electron-hole pair   B. 光电池
3. cavity electron hole   C. 光敏电阻
4. absorption ratio   D. 电子-空穴对
5. visible light   E. 势垒
6. photoelectric triode   F. 空穴
7. multiplier phototube   G. 可见光
8. photoelectric cell   H. 光电倍增管
9. photosensitive resistor   I. 吸收率
10. radiant flux   J. 光电三极管

Ⅱ. Understanding Checking

Directions: Mark Y (for YES) if the statement agrees with the information given in the text; N (for NO) if the statement contradicts the information given in the text; NG (for NOT GIVEN) if the information is not given in the text.

(   )1. A photodetector has a p-n junction that converts a current into light photons.

(   )2. There are a number of performance metrics, also called figures of merit, by which photodetectors are characterized and compared.

(   )3. Dark current is the current flowing through a photodetector in light.

Ⅲ. Passage Completion

Directions: Fill in each of the blanks with one of the words or expressions in the box, making changes if necessary.

| processing | slow | systems | before | wavelengths |

Light can be detected by the eye. The eye is not suitable for modem fiber communications because its response is too __1__, its sensitivity to low-level signals is inadequate, and it is not easily connected to electronic receivers for amplification, decoding, or other signal __2__. Furthermore, the spectral response of the eye is limited to __3__ between 0.4 and 0.7 μm, where fibers have high loss. Nonetheless, the eye is very useful when fibers are tested with visible light. Breaks and discontinuities can be observed by viewing the scattered light. __4__, such as couplers and connectors, can be visually aligned with the visible source __5__ the infrared emitter is attached.

IV. Translation

Directions: Translate the following paragraph from English into Chinese.

The vacuum photodiode and photomultiplier tubes are not placed in operational fiber communications systems, although they can be useful in testing of fiber components. The high sensitivity of the photomultiplier makes it particularly helpful when measuring low levels of optic power. Photoemissive detectors are somewhat easier to explain than semiconductor devices, and the two have many properties in common.

4.3 Optical Detection Techniques

> **Learning objectives:**
> In this section, you will:
> ● understand the definitions of direct detection and coherent detection.
> ● comprehend the definitions of heterodyne detection and homodyne detection.
> ● describe the block diagrams of the detection theory.

Warm-Up Activity

Do you know how these applications work in terms of optical detection techniques?

Picture 1

Picture 2

Picture 3

Picture 4

Picture 5

Text

We will review some basic concepts in optical **detection** for a better understanding of the succeeding study. The most fundamental consideration is that of the photodetector, the device that converts light into an electrical current. There are a variety of detector technologies that show high performance in the visible and infrared portions of the **spectrum** for use by laser radar and communications systems. For reasons of **compactness** and performance, most applications rely on modern semiconductor devices based on the photoelectric effect for the generation of detectable photocurrents. These photocurrents **are comprised of** photo-generated primary electrons and holes in the depletion region of the detector. Modest gains of 10 to 100 can be **achieved** in some detectors through avalanche processes, as in photo-multiplier tubes (PMT) and avalanche photodiodes (APD). These devices typically **exhibit** an excess noise generated by the multiplication or avalanche process, which must be taken into account in the receiver design. Still higher gains can be achieved using APD devices operating in a Geiger mode, whereby the detector is biased well beyond avalanche breakdown, resulting in very high gains ($\sim 10^6$), ultra-fast rise-times (picoseconds), and sensitivity to single-photon events. It has been shown that in all cases, the primary photoelectron **statistics** are **identical** to the Poisson statistics of the **impinging** photon stream, while more complex statistical models are required to describe avalanche processes.

The use of photodetectors in optical receivers can be accomplished in two fundamental ways, either through direct detection or **coherent** detection. Direct detection may be considered a simple energy collection process that only requires a photodetector placed at the focal plane of a lens, followed by an electronic amplifier for signal enhancement. In comparison, coherent detection requires the presence of an optical local oscillator beam to be **mixed with** the signal beam on the photodetector surface. The coherent mixing process **imposes stringent** requirements on signal and local oscillator beam alignment in order to be efficient and can be implemented in two fundamentally different ways. If the signal and local oscillator frequencies are different and **uncorrelated**, the process is **referred to** as **heterodyne** detection, and if they are the same and correlated, as **homodyne** detection. Fig. 4.1 shows a **generic** optical heterodyne configuration in which the signal and local oscillator beams are generated by separate lasers of different, uncorrelated frequencies. They are combined at a beam splitter (BS) that is designed to have a reflectivity high enough to minimize signal loss but low enough to provide sufficient power for use as the local oscillator. Fig. 4.2 shows a possible homodyne arrangement in which a small portion of the transmit beam is used for the local oscillator, thereby satisfying the requirement for correlated frequencies.

generic adj. [生物]属的，类的，一般的

Fig. 4.1 A generic heterodyne detection optical configuration

Fig. 4.2 A generic homodyne optical configuration with a signal Doppler shifted away from baseband by an amount ν_d

Get to Work

I. Matching

Directions: Match the words or expressions in the left column with the Chinese equivalents in the right column.

1. photomultiplier tube (PMT)   A. 多普勒频率
2. beam splitter (BS)   B. 电子(射)束分裂器
3. avalanche photodiode (APD)   C. 光电倍增管
4. Doppler frequency   D. 零差检波器
5. homodyne detector   E. 雪崩光电二极管

II. Understanding Checking

Directions: Mark Y (for YES) if the statement agrees with the information given in the text; N (for NO) if the statement contradicts the information given in the text; NG (for NOT GIVEN) if the information is not given in the text.

(   )1. The most fundamental consideration is that of the photodetector, the device that converts an electrical current into light.

(   )2. The use of photodetectors in optical receivers can be accomplished in two fundamental ways, either through direct detection or coherent detection.

(   )3. Coherent detection requires the presence of an optical local oscillator beam to be mixed with the signal beam on the photodetector surface.

III. Passage Completion

Directions: Fill in each of the blanks with one of the words or expressions in the box, making changes if necessary.

From a system point of view there are several advantages to using homodyne detection over heterodyne detection. __1__, baseband homodyne detection is potentially a factor of 2 more sensitive than heterodyne detection because __2__ the narrower (folded) bandwidth which contains the same amount of energy in half the bandwidth of an unfolded signal, __3__ shown in the figure. Second, frequency-stabilized lasers are usually unnecessary to achieve a narrowband IF in (monostatic) homodyne systems due __4__ the inherent correlation of the transmitted and received frequencies, whereas frequency-stabilized and frequency-locked lasers are usually necessary in heterodyne systems. Third, a single laser system usually offers more compactness and simplicity of design __5__ a two-laser system.

Typical IF spectra from a homodyne system operating at baseband and a heterodyne or offset-homodyne system operating at $\Delta \nu_{IF}$. Note the $\Delta \nu_{IF}$ in bandwidth for the two cases.

IV. Translation

Directions: Translate the following paragraphs from English into Chinese.

In homodyne laser radar applications, correlation between the signal and local oscillator frequencies over the round trip time-of-flight to the target τ_R is usually maintained by using a laser transmitter that has a coherence time that is longer than τ_R. In addition, the transmitter and local oscillator frequencies can be the same or different, depending on whether frequency translators are employed in the optical system. Single-laser frequency offset systems are sometimes referred to as offset-homodyne systems. Frequency offsets can also occur for signals that have been Doppler-shifted away from baseband by an amount ν_d, where $\nu_d = \pm 2|V|/\lambda$ is the Doppler frequency. The negative sign corresponds to motion away from the source so that the Doppler frequency folds over at baseband to produce a positive frequency at ν_d. These processes may still be viewed as homodyne because of the inherent correlation between the transmitted and received frequencies.

Homodyne detection has some unique properties that have proven valuable in the field of quantum optics for demonstrating the possibility of achieving photon noise levels below the quantum limit. Such noise levels FC are referred to as squeezed states, whereby the variance of amplitude or phase, but not both, is reduced below that given by Heisenberg's uncertainty principle for a coherent state. A coherent state is one in which the uncertainty principle is a minimum, otherwise known as laser light. These states are closely related to the bunching and anti-bunching statistics of the photon field.

4.4 CCD

Learning objectives:
In this section, you will:
● understand the structure of a MOS capacitor.
● comprehend the working principle (two phase drive circuit) of CCD analog shift register.

Warm-Up Activity

Do you know the names of the applications based on CCD technology?

Picture 1

Picture 2

Picture 3

Picture 4

Picture 5

Text

A **CCD** (charge-coupled device) is a **MOS** capacitor, the configuration of which is shown in Fig. 4.4. An **electrode** is deposited on a **dielectric film**, formed on the semiconductor substrate.

When a voltage is applied to the electrode, a **depletion layer** is formed in the semiconductor substrate, in the region of the dielectric film. This depletion layer functions as a **pit**, in terms of energy level, for minority carriers. It is thus called a **potential well**. Applying charges to the potential well causes the charges to temporarily accumulate, thus functioning as analog memory. By varying the voltage applied to the electrode, the depth of the well can be adjusted.

A CCD analog shift register consists of a string of CCDs which are used to transfer charges. A CCD **analog shift register** is usually referred to simply as a CCD.

Fig. 4.4 basic CCD structure

CCD Analog Shift Register

This section describes the principle of operation of a CCD analog shift register, taking two-phase driving, illustrated in Fig. 4.5, as an example.

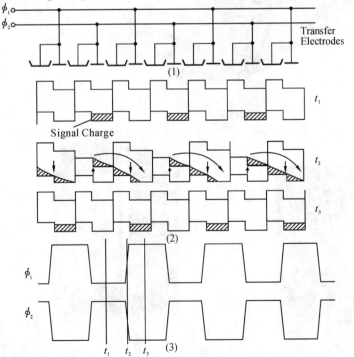

Fig. 4.5 CCD principle of operation (two-phase driving)

CCD 电荷耦合装置，电荷耦合器件
MOS (metal-oxide semiconductor)[计] 金属氧化物半导体
electrode n. 电极
dielectric film 绝缘体薄膜
depletion layer 损耗层
pit n. 深坑，深渊，陷阱
potential well 势垒阱
analog shift register 模拟移位寄存器

Each MOS capacitor consists of two electrodes, under which a potential well having an uneven bottom is formed, thus determining the direction in which charges are transferred. The MOS capacitors are alternately connected to ϕ_1 and ϕ_2.

The signal charge which accumulates under the ϕ_1 electrodes, at time t_1, moves under the adjacent electrodes between times t_2 and t_3. Repeating this operation results in charge transfer.

Get to Work

Ⅰ. Matching

Directions: Match the words or expressions in the left column with the Chinese equivalents in the right column.

1. metal oxide semiconductor (MOS)　　　　A. 图像分割
2. potential barrier　　　　　　　　　　　　B. 二维图像
3. image segmentation　　　　　　　　　　　C. 清晰度
4. two-dimension image　　　　　　　　　　 D. 分辨率
5. resolution　　　　　　　　　　　　　　　E. 像管
6. brightness gain　　　　　　　　　　　　　F. 亮度增益
7. image element　　　　　　　　　　　　　 G. 像素
8. image tube　　　　　　　　　　　　　　　H. 匹配
9. match　　　　　　　　　　　　　　　　　I. 势垒
10. sharpness　　　　　　　　　　　　　　　J. 金属氧化物半导体

Ⅱ. Understanding Checking

Directions: Mark Y (for YES) if the statement agrees with the information given in the text; N (for NO) if the statement contradicts the information given in the text; NG (for NOT GIVEN) if the information is not given in the text.

(　　)1. When a voltage is applied to the electrode, a current is formed in the semiconductor substrate.

(　　)2. A CCD digital shift register consists of a string of CCDs which are used to transfer charges.

(　　)3. By varying the voltage applied to the electrode, the depth of the well can't be adjusted.

(　　)4. The MOS capacitors are alternately connected to ϕ_1 and ϕ_2.

(　　)5. The signal charge which accumulates under the ϕ_1 electrodes, at time t_1, moves under the adjacent electrodes between times t_2 and t_3.

Ⅲ. Passage Completion

Directions: Fill in each of the blanks with one of the words or expressions in the box, making changes if necessary.

charges　　photodiode　　controls　　electrical　　shift

The photosensitive section converts optical energy to 　1　 signals (signal charges) and temporarily accumulates the generated signal __2__ . Fig. 4.6 shows the equivalent circuit of the photosensitive section, which includes a __3__ , MOS capacitor, and transfer gate. The photodiode generates charges, in proportion to the quantity of light and exposure time, which it then accumulates. The transfer gate __4__ the accumulation time (exposure time) and transfers the accumulated signal charges to the CCD analog __5__ register.

Fig. 4.6 Equivalent circuit of the photosensitive section

IV. Translation

Directions: Translate the following from English into Chinese.

Before color CCD linear image sensors became available commercially, color input devices were configured using one of the following methods.

1. Placing a disk having red, green, and blue color filters in front of a monochrome CCD linear image sensor, as a means of performing color separation (color filter switching).

2. Illuminating red, green, and blue light sources, in turn, inputting the produced light to a monochrome CCD linear image sensors (light source switching).

3. Using a prism to separate colors, inputting the three color components to three separate monochrome CCD linear image sensors (prism separation).

4. Mounting color filters on a monochrome CCD linear image sensor chip (on-chip color filter).

CAREER SPEAKING

Making Calls

I. Warm-Up

Model

(A: secretary   B: Henry Green)

A: Good morning, Optical Valley Electronic Components Company.

B: Good morning. Can I speak to Mr. Wu?

A: Let me see if he is **available**. Can I ask who is calling, please?

B: Henry Green from Freedom International Trading Company.

A: Hold the line please. (A few minutes later) Hello, Mr. Green? I'm sorry, but Mr. Wu is in a meeting right now. Can I take a message for you?

B: Yes. Please tell Mr. Wu that Henry Green called. I'm calling about the **contract** my sectary **delivered** to him yesterday morning.

A: Shall I have him return your call?

B: Yes. If he calls me back this afternoon, I'd really appreciate.

A: Does he have your office number and your mobile phone number?

B: I think so, but let me give them to you again.

A: OK.

B: My office number is 027-12345678; my cell phone number is 13900000000.

A: 027-12345678 and 13900000000. Is that right?

B: Yes. He can reach me at my office number before 6:30 p.m. or anytime on my mobile.

A: I see, Mr. Green, and I'll give him your message as soon as Mr. Wu's free.

B: Thank you.

A: You're welcome. Goodbye.

B: Goodbye.

Notes

available	*adj.* 有用的;有空的
contract	*n.* 合同
deliver	*v.* 递交,呈递

Ⅱ. Matching

Directions: Please choose the corresponding sentences from the box to fill in the blanks. Then play roles with your partners.

(A: operator   B: Johnson   C: secretary)

A: Good afternoon, East Lake Wireless Communication Technology Corporation.

_____

B: Good afternoon. This is Johnson from San Francisco. Please connect me with Mr. Brown.

A: _____

C: Hello, Mr. Brown's office.

B: Hello, I'd like to speak to Mr. Brown.

C: I'm sorry. He's not in the office now. _____

B: Yes, thank you. _____ When will he come back?

　　C: I'm not sure. If he is available, I'll call you back. OK?

　　B: OK. Thank you.

　　C: _____

　　B: 695-4426.

> S1: Could you tell me your phone number?
> S2: Hold on and I'll put you through.
> S3: I am phoning to discuss our project with Mr. Brown.
> S4: May I take a message?
> S5: How can I help you?

Ⅲ. Closed Conversation

Directions: Please fill in the blanks according to the Chinese, then play roles with your partner.

　　(A: secretary　B: Kent Black)

　　A: Good morning, East Lake Science and Technology Corporation.

　　B: Good morning. _____(请帮我接你们的经理王先生。)

　　A: _____(请问您是哪位?)

　　B: Kent Black from New York. I called him yesterday.

　　A: Oh, yes, Mr. Black from Universal Trade Corporation. Mr Wang mentioned your name to me. _____(您等一会儿好吗?)

　　(A few minutes later)

　　A: _____
(很抱歉,布莱克先生。王先生现在有客人。您是否要留口信?)

　　B: Yes. _____(请他有空时给我来个电话,好吗?)

　　A: All right. _____
(我会转告他的。能告诉我您的电话吗?)

　　B: 12345678. I'll wait for his call. Thank you.

　　A: It's nothing. Bye.

Ⅳ. Semi-Open Conversation

Directions: Please fill in the blanks with your own words. Then play roles with your partners.

　　(A: Julia　B: Sun Jun　C: Mary)

　　C: Hello, Mary speaking.

　　B: Hello. May I talk to Julia?

C: Sorry, there's no one named Julia here.

B: _____

...

A: Hello. ABC Company. _____

B: Hello. _____

A: This is Julia speaking.

B: Oh, hi, Julia. _____

A: Hi, Sun Jun. I'm glad you called me.

B: You know, Mr. Black will visit your company. I'd like to go over a few details with you.

A: Sorry, _____ But you could talk to Mr. Wang. _____

B: OK.

Ⅴ. Live Show

Challenge 1

Directions: Mr. Zhang calls Mr. White. The secretary answers the phone and tells Mr. Wang that Mr. White is not available now. Please act it out with your partner or partners. The model and the useful expressions below are just for your reference.

Challenge 2

Directions: Lily is calling Mr. Black to discuss the arrangement for meeting a guest. Mr. Black's secretary answers the call and puts Lily through. Please act it out with your partner or partners. The model and the useful expressions below are just for your reference.

Model

(A: operator   B: Zhang Lin   C: secretary   D: Mr. Sharp)

A: Beijing Trading Import and Export Corporation. Good morning. Can I help you?

B: Yes, can you put me through to Mr. Sharp of the Sales Department, please? The **extension** is 335.

A: Could you hold the line while I connect you with Mr. Sharp's office?

B: Yes, thank you.

(A few minutes later)

C: The Sales Department. How may I help you?

B: I'm Zhang Lin calling from Blue Sky Global Corporation. Can I speak to Mr. Sharp?

C: Yes, wait a moment please. (To Mr. Sharp) Mr. Sharp, Mr. Zhang Lin is on the line.

D: Hello, Mr. Zhang. Thanks for calling back. I am **desirous** to know if it's possible to **effect shipment** during October?

B: I'm sorry we can't effect shipment in October.

D: Then, when is the earliest shipment?

B: By the middle of November, I think.

D: That would be too late. I want these goods on our market at the earliest possible to catch the shopping season.

B: Your idea may be a good one, but the trouble is that our manufacturers are fully **committed** at the moment. I'm afraid it'll be difficult for us to **advance** the time for shipment.

D: I hope you'll try to get them to speed up production.

B: All right. We'll do everything to make it.

D: Thanks so much for your help. Bye now.

B: My pleasure. Bye.

Notes

extension	n. 分机
desirous	adj. 渴望的
effect	v. 实现,完成
shipment	n. 装运
commit	v. 承担
advance	v. 提前

Useful expressions

1. 接电话

Hello. ABC Company. May I help you?

This is Mary speaking. Who's that please?

Alice speaking. What can I do for you?

May I know who is calling, please?

2. 打电话

I'd like to speak to Mr. Wang.

Is Mr. Roman in?

I'm trying to get hold of Mr. Chen. Is he available?

Please connect me with Mr. Lee.

Please put me through to your manager.

3. 打错电话

I'm sorry, but I don't know that name.

You must have the wrong number.

There's no one named Lucy here.

Are you sure you have the right name?

I'm afraid you have the wrong number.

4. 没找到人

Mr. Chen is not available at the moment.

He's not in the office now.

He's in Tokyo on business.

He has a visitor right now.

He is in a meeting now.

He's on another line now.

He is engaged now.

5. 等候

Don't hang up. I'll just put you through.

Just a moment, please.

Could you hold a little longer?

Would you mind waiting a few minutes?

Would you care to hold?

Will you hold the line, please?

6. 留口信

Could you take a message, please?

Would you like to leave a message?

I wonder if you could give Mr. Wang a message for me?

Would you like to leave a message on his voice mail, then?

WRITING

Letter of Apology

在工作和生活中,由于疏忽或过失做错了事,给他人造成了麻烦、伤害或损失,最好立即写信给对方赔礼道歉。道歉信通常包括三部分内容:①表示道歉的缘由;②提出补救的措施;③再次道歉,请求谅解。语言要真诚恳切,理由要真实可信,措施要切实可行,切忌掩盖事实、强词夺理。

Sample 1

Dear Mr. Clark,

I am writing this letter to apologize to you for my failing to attend today's meeting.

My wife suddenly had a high fever this morning, and I had to take her to the hospital and stayed there to take care of her. Until my sister arrived, I left and drove back in a hurry. I had thought I could be in time for the meeting but I still missed it. I know the importance of the meeting so I'll ask Sally to let me read the minutes(会议记录) and then I'll write a report to you.

Once again, I apologize for the inconvenience that my absence may cause.

Cordially yours,
Ben

Sample 2

Dear Mr. White,

I would like to give you my apology for the delay in **delivery**. This is due to a thunderstorm that destroyed our plant. Now we're busy **working in three shifts** so as to speed up the shipment. We'll be able to complete the delivery by the end of August. In order to **compensate** your possible loss, we'll give you another 2% **discount**.

I sincerely hope that you can understand our position and accept my apology.

Yours faithfully,
Jerry Morris

Notes

delivery	n. 交付,交货
work in three shifts	三班倒
compensate	v. 补偿
discount	n. 折扣

Useful expressions

1. I would like to give you my apology for...
2. Please allow me to say sorry again.
3. We would appreciate your understanding and ask you to accept our apologies.
4. Please accept my sincere apology for...
5. It's due to my negligence, for which I'm exceedingly sorry.

Challenge

You haven't finished the financial report your manager asked you to do, so you have to write a letter of apology to explain reasons and ask for understanding.

EXTENSIVE READING

4.5 LED Technology

Physical principles

Like a normal diode, the LED consists of a chip of semiconducting material impregnated, or doped, with impurities to create a p-n junction. As in other diodes, current flows easily from the p-side, or anode, to the n-side, or cathode, but not in the reverse direction. Charge-carriers—electrons and holes—flow into the junction from electrodes with different voltages. When an electron meets a hole, it falls into a lower energy level, and releases energy in the form of a photon.

The wavelength of the light emitted, and therefore its color, depends on the band gap energy of the materials forming the p-n junction. In silicon or germanium diodes, the electrons and holes recombine by a non-radiative transition which produces no optical emission, because these are indirect band gap materials. The materials used for the LED have a direct band gap with energies corresponding to near-infrared, visible or near-ultraviolet light.

LED development began with infrared and red devices made with gallium arsenide. Advances in materials science have made possible the production of devices with ever-shorter wavelengths, producing light in a variety of colors.

LEDs are usually built on an n-type substrate, with an electrode attached to the p-type layer deposited on its surface. P-type substrates, while less common, occur as well. Many commercial LEDs, especially GaN/InGaN, also use sapphire substrate.

Colors and Materials

Zinc selenide (ZnSe)—bule.

Gallium(Ⅲ) phosphide (GaP)—red, yellow, green.

Gallium arsenide phosphide (GaAsP)—red, yellow.

Aluminium gallium phosphide (AlGaP)—green.

Indium gallium nitride (InGaN)—Violet, blue.

Aluminium gallium indium phosphide (AlGaInP)—yellow, orange.

Considerations in Use

Unlike incandescent light bulbs, which illuminate regardless of the electrical polarity, LEDs will only light with correct electrical polarity. When the voltage across the p-n junction is in the correct direction, a significant current flows and the device is said to be forward-biased. If the voltage is of the wrong polarity, the device is said to be reverse biased, very little current flows, and no light is emitted. LEDs can be operated on an alternating current voltage, but they will only light with positive voltage, causing the LED to turn on and off at the frequency of the AC supply.

Exercise

Translate the first paragraph into Chinese.

4.6 Optical Storage Technology

Optical storage is a term from engineering, referring to the storage of data on an optically readable medium.

In optical-storage technology, a laser beam encodes digital data onto an optical, or laser, disk in the form of tiny pits arranged in concentric tracks on the disk's surface. A low-power laser scanner is used to "read" these pits, with variations in the intensity of reflected light from the pits being converted into electric signals.

The use of optical storage continues to grow at an incredible pace, spurred on by the flexibility and affordability the technology offers. The word "optical" in the computer industry refers to any storage method which uses a laser to store and retrieve data from media. This term includes such devices as CD-ROM, rewritable optical, WORM (standing for Write-Once-Read-Many), CD-R (short for Compact Disc Recordable), and optical jukeboxes or autochangers.

Most of us are familiar with CD-ROM, but other terms such as rewritable optical, WORM, and CD-R may be foreign. Rewritable optical devices use media that allows data to be written repeatedly, while WORM technology writes data permanently to disk. WORM solutions were initially developed in the late 1970s but never became popular due to the lack of standardization. Today, CD-R and DVD-R (Digital Video Disk Recordable) media have replaced the non-standard devices of the 1970s. Low cost and excellent cross-drive compatibilities make it a popular WORM solution.

CD-R allows up to 650MB of data to be written permanently to a compact disc and read on low-cost CD-ROM drives. These devices are currently sold as single or stand-alone items, or as jukeboxes. Jukeboxes are storage units with built-in robotics to automate access to hundreds of pieces of media and numerous optical drives.

The continuing development of optical technology has opened up many new avenues and created limitless possibilities. Migration software now allows unused data to be moved from its original location to an optical device where it resides until it is needed. Since the only thing that touches the media is the laser, it is the most durable way to store and archive data. Optical storage solutions are also used in a wide variety of applications such as document imaging, records retention, backup systems, desktop publishing, CAD/CAM and many more.

Writable optical, when compared to other random access removable media storage solutions, is the BEST because it has the:

highest capacity available!

lowest cost per megabyte!

longest archive life of any media!

widest environmental condition tolerance!

most reliable media available!

Exercise

Fill in the blanks with the proper words or expressions given below, changing the form if necessary.

| spur | storage | compatibility | CAD | archive |
| random access | automate | encode | built-in | imaging |

1. Unfortunately, this _____ is backwards and so CD players will not play DVDs.
2. Hong Kong is the third-largest film production centre in the world, and the _____ will preserve Hong Kong's rich film heritage.
3. This is a _____ trait of human nature, so do not worry about it.
4. We should _____ the message for security reasons.
5. Primary memory is known as _____ memory and simple named memory.
6. For more than half centuries, the sciences of medical _____ grew steadily but slowly.
7. Few technologies exist to _____ the cognition process.
8. The applicants must be proficient in _____ and other office software.
9. The management accepts full responsibility for loss of goods in _____.
10. A business tax cut is needed to _____ industrial investment.

TRANSLATION

Translation of Attributive Clauses

定语从句的翻译

定语从句是英语中最常用的从句之一,主要分为限制性定语从句和非限制性定语从句。

1. 合成法
合成法就是把定语从句译成位于名词前面的定语词组。
A semi-conductor laser is a special type of solid laser **that uses a semi-conductor**.
半导体激光器是一种使用半导体的特殊种类的固体激光器。

2. 拆分法
如果定语从句很长或者只是对先行词进行补充说明,不太适合翻译成"的"结构的句子,则翻译时可以把复合句拆开,变为一个或几个简短的并列分句。
Analogue displays have a pointer **which moves over a graduated scale**.
模拟显示器有一个指针,该指针可以在刻度盘上移动。

3. 转换法
根据译文表达需要,可将定语从句转换为其他从句或汉语句子的其他成分,如主语、谓语、宾语等。
There is another reason **why the scientists are not in favor of revising the plan**.
为什么科学家们不赞成修改这项计划,还有另外的原因。
(定语转换为主语。)

Challenge

Translate the following sentences into Chinese.

1. Mechanics is that branch of physical science which considers the motion of bodies, with rest as a special case of motion.

2. A technique that has had recent success is laser cooling.

3. The five points which the scientist stressed in his report are very important indeed.

4. It's necessary to consult the tables of technical data which are normally provided in catalogues.

5. The rod is cooled with water, which transmits the pump radiation from the flash lamp or arc lamp.

TASK

Optical Detection

Target	Understand and express professional knowledge.
Requirements	Make an English presentation according to unit 4. Produce a micro-video to express professional knowledge. Add accurate Chinese and English subtitles to the video.
Contents	Optical Detection: Photoelectric effect Photodetector Optical detection techniques

续表

Group：	Optical Detection				
	Project	Name	Software	Time	Score
	Material collection				
	Script				
	PPT				
	Reader				
	Subtitle				
	MP4				
Operating environment	XP/Win7/Win8/ Mobile phone				
Product features					

Script reference

标题	专业内容(讲稿与中、英文字幕)	PPT 中需显示的文字	表现形式	素 材
每个知识点的要点或者小节名字	做报告时的英文讲稿 视频中的中、英文字幕	需要特别提醒的文字	讲解到某个地方或某句话时,出现的图片、动画或者文字	图片或者视频
Contents	**Optical Detection** Contents： Photoelectric effect Photodetector Optical detection techniques **光电探测技术** 内容： 光电效应 光电探测器 光电探测技术	Optical Detection： Photoelectric effect Photodetector Optical detection techniques	文字	

续表

标题	专业内容（讲稿与中、英文字幕）	PPT中需显示的文字	表现形式	素材
1. Photo-electric Effect 光电效应	The photoelectric effect is the name given to the observation that when light is shone onto a piece of metal, a small current flows through the metal. 当光线照在一块金属上时，金属中会有少量电流流过，这个现象称为光电效应。			图1 电子移动（动画图）
	The light is giving its energy to electrons in the atoms of the metal and allowing them to move around, producing the current. 光将其能量传递给金属原子中的电子，并使它们向四周运动，从而产生电流。		动态图	图1 电子移动（动画图）
2. Photo-detectors 光电探测器	The most fundamental consideration is that of the photodetector, the device that converts light into an electrical current. 光电探测器能将光转变为电流。	Photodetector converts light into an electrical current.		图2 光电器件图
	Such as Photosensitive resistor Photoelectric cell Photodiode Photosensitive transistor 如： 光敏电阻 光电池 光电二极管 光电晶体管	Photosensitive resistor Photoelectric cell Photodiode Photosensitive transistor	依次出图显示器件	图2 光电器件图

Unit 4　Optical Detection　　137

续表

标题	专业内容(讲稿与中、英文字幕)	PPT中需显示的文字	表现形式	素　材
Applications 应用举例	Optical detection techniques are widely used, for example: Fingerprint lock Scream application (to measure the crossing sectional area) Rotational speed measurement Illumination meter 光电探测技术应用广泛,如: 指纹锁 工件横截面积测量 转速测量仪 照度计	Fingerprint lock Scream application (to measure the crossing sectional area) Rotational speed measurement Illumination meter	图片示意实例	图3 应用举例
3. Optical Detection Techniques 光电检测技术	There are a variety of detector technologies that show high performance in the visible and infrared portions of the spectrum for use by laser radar and communications systems. 在激光雷达和通信系统中使用了许多探测技术,它们在光谱的红外和可见光部分显示了很高的性能。		显示框图	图4 光电检测系统框图
	For reasons of compactness and performance, most applications rely on modern semiconductor devices based on the photoelectric effect for the generation of detectable photocurrents. 为了简洁和性能方面的要求,大多数应用依赖于"基于光电效应的现代半导体器件"来产生可探测的光电流。			图4 光电检测系统框图
Two fundamental ways 两种基本方式	The use of photodetectors in optical receivers can be accomplished in two fundamental ways, either through direct detection or coherent detection. 在光电接收机上使用光电探测器有两种基本的方式,即直接探测或相干探测。	Direct detection Coherent detection	根据框图进行名词解释	图5 检测技术分类框图

续表

标题	专业内容（讲稿与中、英文字幕）	PPT 中需显示的文字	表现形式	素　材
Direct detection 直接检测	Direct detection may be considered a simple energy collection process that only requires a photodetector placed at the focal plane of a lens, followed by an electronic amplifier for signal enhancement. 直接探测可以认为是一个简单的能量吸收过程,这个过程仅仅需要把一个光电探测器放在透镜的焦面上之后接一个电子放大器对信号放大增强。	Direct detection may be considered a simple energy collection process that only requires a photodetector placed at the focal plane of a lens, followed by an electronic amplifier for signal enhancement.	对照示意图讲解	图6 直接检测图片
Coherent detection 相干检测	In comparison, coherent detection requires the presence of an optical local oscillator beam to be mixed with the signal beam on the photodetector surface. 相比较而言,相干探测在光电探测器的表面要有光学本地振荡器的光束和信号光束混合存在。	In comparison, coherent detection requires the presence of an optical local oscillator beam to be mixed with the signal beam on the photodetector surface.	根据框图进行名词解释,强调 coherent detection	图5 检测技术分类框图
	The coherent mixing process imposes stringent requirements on signal and local oscillator beam alignment in order to be efficient and can be implemented in two fundamentally different ways. 相干混合过程对信号和本机振荡光束的序列强加了严格的要求,以便提高效率,并能够通过两种完全不同的方式实现。	Two fundamentally different ways	强调 coherent detection	图5 检测技术分类框图
	If the signal and local oscillator frequencies are different and uncorrelated, the process is referred to as heterodyne detection, and if they are the same and correlated, as homodyne detection. 如果信号和本机振荡器的频率不同且无关联,则这个过程指的是外差法探测;如果它们的频率相同且相关联,则这个过程就叫零差法探测。	If the signal and local oscillator frequencies are different and uncorrelated, the process is referred to as heterodyne detection, and if they are the same and correlated, as homodyne detection.	对照框图名词解释强调两种方式 heterodyne detection homodyne detection	图5 检测技术分类框图 强调两种方式

续表

标题	专业内容(讲稿与中、英文字幕)	PPT 中需显示的文字	表现形式	素　材
Heterodyne detection 外差检测	Heterodyne detection 外差检测			图 7 heterodyne detection
Heterodyne detection 外差检测	The figure shows a generic optical heterodyne configuration in which the signal and local oscillator beams are generated by separate lasers of different, uncorrelated frequencies. They are combined at a beam splitter (BS) that is designed to have a reflectivity high enough to minimize signal loss but low enough to provide sufficient power for use as the local oscillator. 如图所示为一个通用光学外差结构，其信号和本机振荡光束由两个不同、毫无关联频率的独立激光器产生。它们在分束器处结合。该分束器反射率高足以最大限度减少信号损失,低足以为本机振荡器提供足够的能量。	A generic heterodyne detection optical configuration	对照示意图讲解外差法探测	图 7 heterodyne detection
Homodyne detection 零差检测	Homodyne detection 零差检测			
Homodyne detection 零差检测	The figure shows a possible homodyne arrangement in which a small portion of the transmit beam is used for the local oscillator, thereby satisfying the requirement for correlated frequencies. 如图所示为一种可能的零差配置，其中传输光束的小部分被本机振荡器利用,因此满足关联频率的要求。	A generic homodyne optical configuration with a signal Doppler shifted away from baseband by an amount	对照示意图讲解相干法探测	图 8 homodyne optical configuration

UNIQUE ENGLISH

Exotic Sentences

Directions: Don't rush to make your judgments after the first scanning. Try to understand the true meanings of the following sentences.

1. He is a bicycle doctor.
2. You cannot be too careful.
3. Do you know how to service this new-type machine?
4. He is a good sailor.
5. You are being insulting her!

Unit 5
Laser Processing

In this unit, you are going to learn the following contents:

READING IN DEPTH

5.1 Laser Cutting

5.2 Laser Marking

5.3 Laser Welding

5.4 Laser Perforating

CAREER SPEAKING

After-Sales Service

WRITING

E-mail

EXTENSIVE READING

5.5 Everyday Use of Lasers

5.6 Laser Safety

TRANSLATION

How to Translate with Aid of the Internet and Software

TASK

Laser Marking

UNIQUE ENGLISH

Funny Proverbs

READING IN DEPTH

5.1 Laser Cutting

Learning objectives:

In this module, you will:
- understand the concept of laser cutting.
- learn about types of laser cutting.
- master differences between laser cutting and mechanical cutting.

Warm-Up Activity

Describe the following pictures.

Words and phrases for reference: fall in love with, marriage proposal, diamond ring, come up with an idea, jeweler, laser marking

Picture 1

Picture 2

Picture 3

Picture 4

Picture 5

Text

Laser cutting is a technology that uses a laser to cut materials, and is usually used in industrial manufacturing. Laser cutting works by **directing** the output of a high power laser, by computer, **at** the material to be cut. The material then either **melts**, burns or **vaporizes** away leaving an edge with a high quality surface finish. Industrial laser cutters are used to cut **flat-sheet material** as well as structural and piping materials. Some 6-**axis** lasers can perform cutting operations on parts that have been pre-formed by **casting** or **machining**.

Comparison to Mechanical Cutting

Advantages of laser cutting over mechanical cutting **vary** according to the situation, but two important factors are the lack of physical contact (since there is no cutting edge which can become **contaminated** by the material or contaminate the material), and to some extent **precision** (since there is no **wear** on the laser). There is also a

direct at 把……对准
melt v. （使）融化，（使）熔化，使软化
vaporize v. （使）蒸发
flat-sheet material 平板材料
axis n. 轴
cast v. 铸造
machine vt. 机器制造
vary vi. 变化，不同
contaminate v. 污染
precision n. 精确，精度
wear n. 磨损

reduced chance of **warping** the material that is being cut as laser systems have a small heat affected zone. Some materials are also very difficult or impossible to cut by more traditional means. One of the disadvantages of laser cutting may include the high energy required.

Types

The most popular lasers for cutting materials are **CO_2** and Nd：YAG, though semiconductor lasers are gaining **prominence** due to higher efficiency. Typically there is a choice between a DC (direct current) and RF (radio frequency) powered resonator (laser generator). The fundamental choice of beam generation method can have significant impact on productivity and life cycle costs due to the differences between the two types. DC resonators have internal electrodes **encapsulated** in the cavity glass where RF resonators have external electrodes.

In addition to the power source, the type of **gas flow** can affect performance as well. In a fast **axial flow** resonator, the mixture of carbon dioxide, **helium** and **nitrogen** is circulated at high velocity by a turbine slab or blower. **Diffusion** cooled resonators have a static gas field that requires no **pressurization** or glassware, leading to savings on replacement of turbines and glassware.

Process

Laser cutters usually work much like a **milling machine** for working a sheet in that the laser (equivalent to the mill) enters through the side of the sheet and cuts it through the axis of the beam. In order to be able to start cutting from somewhere else than the edge, a **pierce** is done before every cut. Piercing usually involves a high power pulsed laser beam which slowly (taking around 5-15 s for half-inch thick stainless steel, for example) makes a hole in the material.

Get to Work

Ⅰ. Matching

Directions: *Match the words or expressions in the left column with the Chinese equivalents in the right column.*

Unit 5 Laser Processing   145

1. high power diode laser   A. 切割质量
2. supersonic nozzle   B. 实时监测
3. jet flow   C. 光信号
4. dynamic characteristic   D. 精密切割
5. real-time monitoring   E. 射流场
6. optical signal   F. 激光技术
7. cutting quality   G. 按键面板
8. laser technique   H. 动力学特性
9. keystroke   I. 超音速喷嘴
10. precision cutting   J. 大功率半导体激光器

Ⅱ. **Understanding Checking**

Directions: Fill in the missing parts of the short passage with no more than 3 words based on the above text.

Laser cutting can make material melt, burn and _____ leaving an edge with a high quality surface finish.

Industrial laser cutters are used to cut piping materials, structural materials and _____.

Laser cutting has many advantages over mechanical cutting, such as lack of _____, precision and less deformation.

Even if semiconductor laser is excellent for its _____, the most popular lasers for cutting materials are CO_2 and Nd:YAG.

To start cutting form somewhere else than the edge, _____ should be done before every cut.

Ⅲ. **Passage Completion**

Directions: Fill in each of the blanks with one of the words or expressions in the box, making changes if necessary.

| availability | offer | processing | reliable | speed |
| identical | due to | advantage | desirable | accomplish |

When high-strength steel is concerned, the industry is increasingly turning to lasers. There are some reasons.

Firstly, in contrast to mechanical cutting, more can __1__ with light. The laser being an optical, non-contact tool has big __2__, especially since the melting point of "usual" and high-strength steel is almost __3__.

Second, short cycle times are especially __4__. One disadvantage that laser cutting has compared to mechanical punching is its low __5__. But that can be increased. Prime examples are the 3D laser __6__ machines of the TruLaser Cell Series 7000. They __7__ sensor-supported piercing on-the-fly and plasma high-speed cutting. Overall, the laser system provides increased machine power and added dynamics __8__ the decreased moving masses.

Thirdly, another significant factor is a high __9__. The laser cutting systems, which have proven themselves in practical applications a thousand times over, are very __10__.

Ⅳ. Phrase Translation

Directions: Complete the sentences by translating into English according to the Chinese given in brackets.

1. The input signal modulates the drive current, so the output optical signal _____ (与输入电信号成比例).

2. The trunk lines that connect central offices _____ (已被光纤所取代).

3. Fiber optic cables _____ (对电磁噪音具有很强的抵御能力) such as radios, motors or other nearby cables.

4. _____ (光纤现在已经成为一个主要部分) of the infrastructure for a national telecommunication information highway in the U.S. and all around the world.

5. A low-power laser scanner is used to "read" these pits, with variations in the intensity of reflected light from the pits _____ (被转化为电子信号).

Ⅴ. Sentence Remaking

Directions: Simulate the following sentence patterns according to an example provided from the text.

1. ...is/are...that uses...to...

E.g. Laser cutting is a technology that uses a laser to cut materials.

_____ is _____ that uses _____ to _____.

2. ...is/are used to...as well as...

E.g. Industrial laser cutters are used to cut flat-sheet material as well as structural and piping materials.

_____ is/are used to _____ as well as _____.

3. ...is/are very difficult or impossible to...

E.g. Some materials are very difficult or impossible to cut by more traditional means.

_____ are very difficult or impossible to _____.

4. In addition to..., ...can...as well.

E.g. In addition to the power source, the type of gas flow can affect performance as well.

In addition to _____, _____.

can _____ as well.

5. In order to..., ...is/are done before...

E. g. In order to be able to start cutting from somewhere else than the edge, a pierce is done before every cut.

In order to _____, _____ is done before _____.

VI. Para. Translation

In 1965, the first production laser cutting machine was used to drill holes in diamond dies (金刚石拉模). This machine was made by the Western Electric Engineering Research Center. In 1967, the British pioneered laser-assisted oxygen jet cutting (射流切割) for metals. In the early 1970s, this technology was put into production to cut titanium (钛) for aerospace applications. At the same time, CO_2 lasers were adapted to cut non-metals, such as textiles, because, at the time, CO_2 lasers were not powerful enough to overcome the thermal conductivity (热导率) of metals.

VII. Mini Imitative Writing

Directions: The following paragraph originates from the text. Please imitate it and write another one.

Advantages of laser cutting over mechanical cutting vary according to the situation, but two important factors are the lack of physical contact, and to some extent precision. One of the disadvantages of laser cutting may include the high energy required.

5.2 Laser Marking

> **Learning objectives:**
> In this module, you will:
> ● understand the concept of laser marking.
> ● learn about principles of laser marking.
> ● master differences between laser marking and mechanical marking.

Warm-Up Activity

Look at the picture, and try to match the English phrases with the Chinese equivalents.

1. ignition box
2. Q-switch
3. krypton arc lamps
4. mode aperture
5. laser rod
6. shutter
7. rear mirror
8. front mirror
9. beam expander
10. laser diode
11. galvo head
12. flat field lens
13. marking field

A. 激光棒
B. 激光二极管
C. 点火盒
D. 调 Q 开关
E. 后反射镜
F. 氪灯
G. 准直扩束镜
H. 打标区
I. 快门
J. 场镜
K. 选模小孔光阑
L. 前镜
M. 振镜

Text

Laser marking is a non-contact thermal process that uses the heat generated by a laser beam to **alter** the surface of the workpiece. The laser light must be absorbed by the material surface to generate the required heat for marking. At present, there are two types recognized principles: by hot working and by cool working.

Hot working uses laser beam with higher energy **density** to **irradiate** on the materials, the materials become hotter by absorbing the laser energy and with phenomena of melting, **ablation** and **evaporation**, etc.

Cool working uses high road energy ultraviolet photons to break materials (especially **organic** materials) or chemical bonds in the mediums around, to destroy materials in the non-hot process. Cool working processing has special **significance** for laser marking, because it does not work through hot working, and will not produce hot damage or break cold **stripping** of chemical bonds, therefore, no hot **deformation** will be caused on the surface of the materials.

Compared with other marking technologies such as ink jet printing and mechanical marking, laser marking has a number of advantages, such as very high processing speeds, low operation cost (no use of **consumables**), constant high quality and **durability** of the results, avoiding **contaminations**, the ability to write very small features, and very high **flexibility** in automation.

The physical principle of laser marking is based on *Light Amplification by Simulated Emission of Radiation*. In other words, the energy emitted by a source (e.g. arc lamp) onto an active source (e.g. Nd:Yag crystal) will be collected and concentrated between two opposing mirrors.

alter *v.* 改变

density *n.* 密度
irradiate *v.* 照射
ablation *n.* 消融,切除
evaporation *n.* 蒸发(作用)
organic *adj.* 有机的,组织的
significance *n.* 意义,重要性
stripping *n.* 抽锭,脱模
deformation *n.* 变形
consumable *n.* 耗材
durability *n.* 经久,耐久力
contamination *n.* 玷污,污染,污染物
flexibility *n.* 弹性,适应性

Scanning this spot on the part produces marking, either with material removal or reaction at the surface of the part.

1. The oscillation of specific light particles (photons) that move between the mirrors generates a high power laser beam. Only part of this beam passes through the output mirror (typically 10% to 20%), the rest being **reinjected** for the amplification process; the laser beam is thus created.

2. The beam is **deflected** by two scanning mirrors, driven electronically and via software, thus enabling a very high scanning speed and **exceptional** positioning **accuracy**.

3. The flat **field lens** then concentrates the beam energy in a very tiny spot and greatly increases the power density on the surface to be marked.

The different markings can be classified in three categories.

Engraving with Material Removal (sublimation)

The energy is **delivered** with high peak power pulses so that the material is instantaneously removed without thermal side-effects on the parts.

Material Annealing with Surface Color Change

The energy is delivered with lower pulses, heating the material and changing the surface appearance. No material is removed.

Marking Through Layer Removal

On **coated** material, the contrast is created by removing the top layer, thus showing the color of the base material.

reinject *vt.* 再注入,给……作再注射
deflect *v.* (使)偏斜,(使)偏转
exceptional *adj.* 例外的,异常的
accuracy *n.* 精确性,正确度
field lens 物镜
sublimation *n.* 升华
deliver *vt.* 递送,释放

anneal *n.* 退火

coated *adj.* 有涂层的

Get to Work

I. Matching

Directions: Match the words or expressions in the left column with the Chinese equivalents in the right column.

1. unit amplitude   A. 平均色散
2. tracking lidar   B. 化学键
3. magnifying glass   C. 平行反射镜
4. mean dispersion   D. 单位振幅
5. chemical bond   E. 反射密度
6. closed resonator   F. 有机半导体
7. parallel mirror   G. 跟踪激光雷达
8. zero deflection   H. 闭合共振器
9. reflection density   I. 放大镜
10. organic semiconductor   J. 零偏转,无偏转

Ⅱ. Understanding Checking

Directions: Fill in the missing parts with no more than 3 words based on the above text.

1. Laser marking is a _____ thermal process that uses the heat generated by a laser beam to alter the surface of the workpiece.

2. There are two types principles: _____ and _____.

3. Hot working makes the materials become hotter by _____ the laser energy and with the phenomena of melting, ablation and evaporation.

4. Cool working use _____ to destroy materials in the non-hot process.

5. Laser marking has many advantages over mechanical marking, such as high speed, low cost, high quality and _____ of the result.

Ⅲ. Passage Completion

Directions: Fill in each of the blanks with one of the words or expressions in the box, making changes if necessary.

| fiber-optics | signal | convert | modulate | distorted |
| copper wire | advantage | revolutionize | transmitter | role |

Fiber-optic communication is a method of transmitting information from one place to another by sending light through an optical fiber. The light forms an electromagnetic carrier wave that is __1__ to carry information. First developed in the 1970s, fiber-optic communication systems have __2__ the telecommunications industry and played a major __3__ in the advent of the information age. Because of its __4__ over electrical transmission, the use of optical fiber has largely replaced __5__ communications in core networks in the developed world.

The process of communicating using __6__ involves the following basic steps: creating the optical signal using a __7__, relaying the __8__ along the fiber, ensuring that the signal does not become too __9__ or weak, and receiving the optical signal and __10__ it into an electrical signal.

Ⅳ. Translation

Directions: Complete the sentences by translating into English according to the Chinese given in brackets.

1. The external modulator can change beam intensity _____ (不会直接影响激光的工作状态), avoiding chirp or any wavelength drift caused by pulsing the laser.

2. An ideal transmission medium would _____ (对它所传输的信号不产生影响), but any medium inevitably has some effects.

3. Both attenuation and dispersion _____ （随波长的改变而改变）, so fibers have certain transmission windows.

4. Pulsed lasers which provide _____ （高冲击能）for a short period are very effective in some laser cutting processes, particularly for piercing.

5. In optical-storage technology, a laser beam encodes digital data onto an optical disk in the form of tiny pits arranged _____ （以同心轨道方式）on the disk's surface.

Ⅴ. Sentence Remaking

Directions: Simulate the following sentence patterns according to an example provided from the text.

1. ... have special significance for...

E.g. Cool working processing has special significance for laser marking.

_____ has special significance for _____ .

2. Compared with..., ... have a number of advantages.

E.g. Compared with other marking technologies, laser marking has a number of advantages.

Compared with _____ , _____ has a number of advantages.

3. ... is/are based on...

E.g. The physical principle of laser marking is based on *Light Amplification by Simulated Emission of Radiation*.

_____ is based on _____ .

4. ... is/are created by..., thus...

E.g. On coated material, the contrast is created by removing the top layer, thus showing the color of the base material.

_____ is created by _____ , thus _____ .

5. ... be absorbed by...

E.g. The laser light must be absorbed by the material surface to generate the required heat for marking.

_____ be absorbed by _____ .

Ⅵ. Para. Translation

Markings are required on workpieces（工件）in all branches of industry. The attachment of numbers, texts or identity codes has become a part of the added value process. The laser has established itself as a progressive marking tools for all these tasks because it fulfills the requirements to a special degree. The laser beam（激光束）allows non-contact marking and is free of wear and tear（磨损）, making it ideal for marking any product shapes, even at hard-to-reach locations.

Ⅶ. Mini imitative writing

Directions: The following paragraph originates from the text. Please imitate it and write another one.

Compared with other marking technologies such as ink jet printing and mechanical marking, laser marking has a number of advantages, such as very high processing speeds, low operation cost, constant high quality and durability of the results, avoiding contaminations, the ability to write very small features, and very high flexibility in automation.

5.3 Laser Welding

Learning objectives:

In this module, you will:
- understand the features of laser welding.
- learn about two main processes: heat conduction welding and deep penetration welding.
- master the advantages of laser welding.

Warm-Up Activity

Describe the following pictures.

To Lose One's Hair
Picture 1

Hair Transplant Laser
Picture 2

Laser Hair Transplant
Picture 3

New Look
Picture 4

Job Interview
Picture 5

Success
Picture 6

Text

The laser beam welding is mainly used for joining **components** that need to be joined with high welding speeds, thin and small weld **seams** and low thermal **distortion**. The high welding speeds, an excellent automatic operation and the possibility to control the quality online during the process make the laser welding a common joining method in the modern industrial production.

component *n.* 成分

seam *n.* 接缝，线缝
distortion *n.* 扭曲，变形

In laser welding one must **distinguish** between two main processes: heat **conduction** welding and deep **penetration** welding.

In heat conduction welding, the materials to be joined are melted by **absorption** of the laser beam at material surface—the **solidified** melt joins the materials. Welding penetration depths in this context are typically below 2 mm.

Deep penetration welding, which starts at energy densities of **approx.** 106 W/cm², is based on the creation of a vapor **capillary** inside the material by local heating to the evaporation temperature. The resulting vapor pressure inside the material creates a capillary approx. 1.5 times the diameter of the **focal spot** of the laser beam, which is moved through the material by the device, following the **contour** to be welded. The **hydrostatic** pressure, the surface tension of the melt, and the vapor pressure inside the capillary reach a balance, preventing the capillary (often referred to as the "**keyhole**") from **collapsing**. Multiple reflection inside the keyhole guides the incident laser beam deep into the material. Today, given sufficient laser power, weld depths of up to 25 mm (steel) can be achieved.

Heat Conduction Welding   Deep Penetration Welding

① Plasma Cloud
② Molten Material
③ Keyhole
④ Weld Depth

Compared to **conventional** welding methods, laser welding offers the **diverse** advantages: no tool wear, contact-free processing, diverse materials, and high welding seam quality. **Provided** the two parts to be joined is good (i.e. a small gap relative to the thickness of material), no additional material needs to be added, such as a **welding rod**. There are occasions however when it is **desirable** to add material, for example:

When **dissimilar** materials cannot be welded together to **yield** a strong joint, **metallurgy** can point to a suitable **intermediate** material which can be added to the weld in rod form to make a **crackfree** joint possible. In tool repair, **worn** surfaces can be built up so that tools can be **refurbished**. In jewellery repair, it is often necessary to add material to effect repairs such as **reprocessing** jewel mounts or **resizing** rings. Laser welding with fine wire of the same material as the original item makes this possible.

metallurgy *n.* 冶金，冶金术
intermediate *adj.* 中间的 *n.* 媒介
crackfree *adj.* 无裂缝的
worn *adj.* 用旧的，疲倦的
refurbish *vt.* 再磨光，刷新
reprocess *v.* 再生，再加工
resize *v.* 调整大小

Get to Work

Ⅰ. Matching

Directions: *Match the words or expressions in the left column with the Chinese equivalents in the right column.*

1. photoreader
2. photojunction diode
3. adaptive filter
4. three-dimensional laser processing
5. all-purpose instrument
6. image sensor
7. discharge tube
8. passive imaging system
9. absorption coefficient
10. optical memory

A. 吸收系数
B. 自适应滤光片
C. 三维激光加工
D. 光学存储器
E. 光电读出器
F. 被动成像系统
G. 图像传感器
H. 通用仪器
I. 光电二极管
J. 放电管

Ⅱ. Understanding Checking

Directions: *Mark Y (for YES) if the statement agrees with the information given in the text; N (for NO) if the statement contradicts the information given in the text; NG (for NOT GIVEN) if the information is not given in the text.*

(　　)1. Laser welding is widely used for joining components because of its high welding speeds and low thermal distortion.

(　　)2. There are two main processes of laser welding, that is, heat welding and cool welding.

(　　)3. Deep penetration welding depths can range from 2 mm to 25 mm.

(　　)4. If the two parts to be joined have a small gap, a welding rod needs to be added.

(　　)5. When dissimilar materials cannot be welded together, a suitable intermediate material can be added to the weld to make a crackfree joint possible.

Ⅲ. Passage Completion

Directions: *Fill in each of the blanks with one of the words or expressions in the box, making changes if necessary.*

| announce | abandon | replace | all-around | single standard |
| as of | support | while | storage format | higher-quality |

The beginnings of the DVD format may be traced back to the 1990s, where two optical __1__ were in the works. Philips and Sony had the MultiMedia Compact Disc (MMCD), __2__ numerous companies including Toshiba, Time-Warner, Matsushita Electric, Mitsubishi Electric, Pioneer, Thomson, and JVC __3__ the Super Density Disc (SDD). However, then-president of IBM Lou Gerstner moved for a __4__ and a unification of two camps. The MMCD __5__ and the SD format took on a number of revisions. In 1995, the DVD specification Version 1.5 __6__ . By 1997, the DVD Forum __7__ the DVD Consortium (联盟) and was opened to all companies.

__8__ today there is no official meaning behind the letters DVD. The term "Digital Versatile Disc" came first, as the primary purpose of the DVD was to fully realize a __9__ home theater experience. Some members of the DVD Forum, however, believe "Digital Versatile Disc" is more appropriate to denote __10__ use and not just video.

Ⅳ. **Translation**

Directions: Complete the sentences by translating into English according to the Chinese given in brackets.

1. With sufficiently high energy density, the laser beam heats the material to such a point, that it is _____ (不仅在表面熔化,而且蒸发).

2. In military and defense, _____ (快速传递大量信息) is the impetus behind a wide range of retrofit and new fiber optic programs.

3. It was Bell himself who _____ (发明了最早的光波通信设备) in 1880.

4. This location, called a node, would provide the optical receiver that _____ (将光脉冲转化为电信号).

5. In a fiber, the two principal limiting effects are attenuation of the signal strength and _____ (脉冲的色散).

Ⅴ. **Sentence Remaking**

Directions: Simulate the following sentence patterns according to an example provided from the text.

1. …make…a…in…

E.g. The high welding speeds make the laser welding a common joining method in the modern industrial production.

_____ make _____ a _____ in _____ .

2. ..., which..., is based on...

E. g. Deep penetration welding, which starts at energy densities of approx. 106 W/cm^2, is based on the creation of a vapor capillary inside the material.

_____, which _____, is based on _____.

3. Given..., ... can be achieved.

E. g. Given sufficient laser power, weld depths of up to 25 mm (steel) can be achieved.

Given _____, _____ can be achieved.

4. Provided..., no... need...

E. g. Provided the two parts to be joined is good (i. e. a small gap relative to the thickness of material), no additional material needs to be added.

Provided _____, no _____ need _____.

5. ...make...possible.

E. g. Laser welding with fine wire of the same material as the original item makes this possible.

_____ make _____ possible.

VI. Para. Translation

Directions: Translate the following paragraph into Chinese.

The specific features of manual welding systems have revolutionized manufacturing and repair work in the jewelry industry. Work pieces are manually positioned in a working chamber and viewed through a microscope (显微镜) with crosshair (十字瞄准线). They are permanently spot welded making expensive clamps (夹钳) or time-consuming wire binding unnecessary. Precious metals and alloys (合金) can be welded without any filler (填装物) or soldering material (焊接材料). Laser weld seam is extremely robust and requires only little post processing. Impurities (杂质) of potentially toxic (有毒的) alloys are avoided.

VII. Mini Imitative Writing

Directions: The following paragraph originates from the text. Please imitate it and write another one. Define a device or technique related to laser.

The laser beam welding is mainly used for joining components that need to be joined with high welding speeds, thin and small weld seams and low thermal distortion. The high welding speeds, an excellent automatic operation and the possibility to control the quality online during the process make the laser welding a common joining method in the modern industrial production.

5.4 Laser Perforating

Learning objectives:

In this module, you will:
● understand the purpose of laser perforating.
● learn about applications of laser perforating.
● know the process of how to make "easy open" packaging.

Warm-Up Activity

Please write an individual signature (个性签名) for each of the following functions of laser perforation. You may discuss with your partner with help of Picture 1 as an example. Some of you will be invited to show your created individual signatures to the whole class.

Cigarette Tipping Paper
Picture 1

Individual signature: I'm the mascot of smokers!

Atmosphere Packaging
Picture 2

Individual signature: _____
_____.

Easy Open Packaging
Picture 3

Anti-Forgery Passport
Picture 4

Individual signature：_____
_____.

Individual signature：_____
_____.

Text

Lasers are used to drill tiny holes at high speed in webs of thin paper and plastic. The purpose of these **perforations** is generally to create weaknesses to allow packages to open in a **predefined** way ("Easy Open") or to make the material **porous** to allow air or steam to move through the barrier in a controlled way.

perforation n. 穿孔
predefine vt. 预先确定
porous adj. 多孔渗水的

Multiple laser heads perforate plastic and paper in-line

Application Examples
Perforation of Cigarette Tipping Paper

By producing **an array of** holes in the range of 50 μm to 100 μm in tipping paper, cigarette manufacturers reduce the **tar inhaled** by the smoker as cool air is drawn into the **filter**, which encourages the tar vapour to **condense** inside the filter, rather than being inhaled into the **lungs**.

A kilowatt CO_2 laser is split into multiple beams and delivered to the paper by means of **stationary** microfocusing heads. In the most rapid winding systems, such systems are capable of perforating at approaching 500 000 holes per second!

Laser Perforation of Modified Atmosphere Packaging

As cut fruit and vegetables continue to **respire** within plastic

tipping paper 接装纸
an array of 一排，一群，一批
tar n. 焦油，柏油
inhale vt. 吸入
filter n. 滤波器，过滤器
condense v. （使）浓缩，精简
lung n. 肺，呼吸器
stationary adj. 固定的
modified adj. 改良的，改进的，修正的
respire v. 呼吸

packaging, the **shelf life** of produce can be extended by making the packaging "**breathable**"—i. e. **permeable** to oxygen and carbon dioxide. Laser-drilled holes reliably create the correct size and density of holes to **facilitate** this without **drying out** the produce.

"Easy Open" Packaging

Lasers can **scribe** or perforate materials which will subsequently be used to package candies, snacks, coffee, and drinks to make the product easier to access. Tear lines can be created by scribing or making a **linear** perforation to weaken the packaging where it is required. The CO_2 laser is well absorbed by organic materials but not by foil barriers used in peanut and crisp packs meaning that the contents remain **sealed** from atmosphere.

Similarly, entry holes for **drinking straws** and exit areas for **blister** packs can be made so much easier to use by selective package weakening. Creating such structure on a web of material whilst it moves between **rollers** can require high speed movement of the laser spot across the web. Generally these applications therefore require "**galvo**" optics which deflect the laser at very high speed using rotating pairs of mirrors.

Anti-forgery Ticket and Passport Perforation

Passports can have the number cut through all the pages by **steering** the laser to create **alphanumerics** made from a **matrix** of circles. World cup tickets have been finished with **intricate tear-off** perforations which can be very difficult for **forgers** to **emulate**.

shelf life (包装食品的)货架期,保存限期
breathable *adj.* 通气的,可渗透的
permeable *adj.* 有浸透性的,能透过的

facilitate *vt.* 使便利,促进
dry out 干透,使干
scribe *vt.* 用划线器划
linear *adj.* 线的,直线的,线性的
sealed *adj.* 未知的,密封的
drinking straw 吸管
blister *n.* 水泡
roller *n.* 滚筒,辊子
galvo *n.* 检流计
anti-forgery 防伪
steer *v.* 驾驶,掌舵
alphanumeric *n.* 字母和数字组合
matrix *n.* 矩阵
intricate *adj.* 复杂的,错综的
tear-off *adj.* 可按虚线撕开的
forger *n.* 伪造者
emulate *v.* 仿效

Get to Work

I. Matching

Directions: *Match the words or expressions in the left column with the Chinese equivalents in the right column.*

1. electro-optics
2. light-wave
3. trunk line
4. milling machine
5. beam divergence
6. transverse mode
7. thermal lensing
8. infrared spectrum
9. output coupling
10. doping concentration

A. 射束发散性
B. 红外线谱
C. 电光学
D. 输出耦合
E. 横模,横向模态
F. 光波
G. 干线,中继线
H. 掺杂浓度
I. 铣床
J. 热透镜效应

Ⅱ. Understanding Checking

Directions: *Choose the best answer for each statement below according to the text.*

1. The purpose of laser perforations is generally to create _____ to allow package easy open.
　　A. strength　　B. energy source　　C. weakness　　D. barrier

2. Laser perforation can make fruit and vegetables continue to _____ within plastic packaging.
　　A. grow　　B. extend　　C. respire　　D. breathe

3. "Easy open" packaging is often used to package the following except _____.
　　A. snacks　　B. sweetmeat　　C. juice　　D. cigarette tipping paper

4. Galvo optics _____ the laser at very high speed using rotating pairs of mirrors.
　　A. radiate　　B. deflect　　C. convey　　D. distort

5. The word "alphanumerics" in the second line of last paragraph means _____.
　　A. letters and numbers　　B. alphabets　　C. pictures　　D. figures

Ⅲ. Passage Completion

Directions: *Fill in each of the blanks with one of the words or expressions in the box, making changes if necessary.*

| develop | prospect | appropriate | combine with | efficiency |
| non-metallic | widely | promising | extremely | suitable |

　　Lasers are new light sources in 1960s. They are __1__ used due to lasers' good direction, high brightness. Laser processing is one of the most __2__ fields in laser application, now more than 20 kinds of laser processing technology have been __3__.

　　Lasers' space and time control is very good. There are very large spaces for the material of the object, shapes, sizes and processing conditions, especially __4__ for automation processing. Laser processing system __5__ computer numerical control technology can form a high efficient automatic processing equipment which has become a key technology for the enterprise's production. It has opened up a broad __6__ for high quality, high __7__ and low cost manufacturing.

　　Both hot processing and cold processing can be used in metallic and __8__ materials, like cutting, drilling, marking. Hot processing is __9__ beneficial for metal materials welding, surface treatment, alloy production, cutting. Cold processing is __10__ for light chemical deposition, rapid laser prototyping technology, laser etching, mixed dyeing and oxidation.

Ⅳ. Translation

Directions: *Complete the sentences by translating into English according to the Chinese given in brackets.*

1. The advent of lasers in 1960 and low-loss optical fiber in 1970 _____

_____ (消除了所有这些障碍).

2. In recent years it has become apparent that _____ (光纤正一步步替代铜线) as an appropriate means of communication signal transmission.

3. Since _____ (唯一接触到媒介的物质) is the laser, it is the most durable way to store and archive data.

4. _____ (光学技术的持续发展) has opened up many new avenues and created limitless possibilities.

5. Jukeboxes are storage units _____ (有内置机器人技术) to automate access to hundreds of pieces of media and numerous optical drives.

Ⅴ. Sentence Remaking

Directions: Simulate the following sentence patterns according to an example provided from the text.

1. The purpose of... is generally to...

E. g. The purpose of these perforations is generally to create weaknesses to allow packages to open in a predefined way.

The purpose of _____ is generally to _____ .

2. ...is...but not...

E. g. The CO_2 laser is well absorbed by organic materials but not by foil barriers used in peanut and crisp packs.

_____ is _____ but not _____ .

3. As..., ...can be extended.

E. g. As cut fruit and vegetables continue to respire within plastic packaging, the shelf life of produce can be extended.

As _____ , _____ can be extended.

4. Similarly, ...can be made...

E. g. Similarly, entry holes for drinking straws and exit areas for blister packs can be made so much easier to use by selective package weakening.

Similarly, _____ can be made _____ .

5. ...have been finished with...

E. g. World cup tickets have been finished with intricate tear-off perforations.

_____ have been finished with _____ .

Ⅵ. Para. Translation

Directions: Translate the following paragraphs into Chinese.

1. Cigarette-tip（烟嘴）paper is perforated mainly to reduce the tar and nicotine content of cigarettes. Laser perforation ensures the definite adherence to the threshold values（界限值）.

2. The laser perforation of packaging films is used to prolong the freshness and quality of perishable（易腐坏的）food. Micro perforation enhances the shelf life of vegetables etc. by exchanging oxygen through micro holes in the range of 60-100 μm.

3. The individual perforation of identification documents or VIP entrance cards with CO_2 lasers guarantees to gain maximum protection against forgery.

Ⅶ. Mini Imitative Writing

Directions：The following paragraph originates from the text. Please imitate it and write another one. Describe a device's function.

Lasers can scribe or perforate materials which will subsequently be used to package candies, snacks, coffee, and drinks to make the product easier to access. Tear lines can be created by scribing or making a linear perforation to weaken the packaging where it is required.

CAREER SPEAKING

After-Sales Service

Ⅰ. Warm-Up

Model

(A：customer   B：receptionist)

A：I bought an **air-conditioner** at a supermarket. But it had only been used for one week before it couldn't work.

B：I suggest you ask that supermarket to repair or replace it. They are responsible for providing after-sales service.

A：Yes, I did. But the manager of the supermarket told me they had sold out all the air-conditioners with that model. What's worse, they were short of some **spare parts**. So they asked me to turn to you for help.

B：Could you tell me what your model number is?

A：PIU-9254.

B: All right. Our worker will be sent to your place tomorrow morning if it's convenient to you.

A: Oh, unfortunately, I'll go abroad on business tomorrow. How about this afternoon? I've asked for leave to stay at home.

B: Let me check whether there is someone available. You know, most of our workers have gone out to deliver goods.

A: I'm sorry to have given you so much trouble.

B: It doesn't matter, sir. Well, Paul is free. He'll get there two hours later. Please wait for him at home.

A: OK, thank you.

B: You're welcome. Is there anything else I can help you?

A: No, I'm satisfied with your service.

B: Thanks for your support.

Notes

air-conditioner *n.* 空调

spare part 备件

II. Matching

Directions: Please choose the corresponding sentences from the box to fill in the blanks. Then play roles with your partner.

(A: assistant manager   B: customer)

A: Hello, this is Customer Service Department of Grace Electrical Equipment Company. _____

B: Hello, I ordered 3,000 TV sets from your company last month, but I find 20 of them can't work.

A: I'm sorry to hear that. _____

B: Yes, we've already sent them.

A: _____

B: All right. How long will it take?

A: _____ I feel sorry again.

S1: If the responsibility actually lies in us, we will send you a replacement right away and compensate your loss.

S2: Would you like to ship back the faulty ones in case that we should make a through investigation?

S3: May I help you?

S4: Almost one week after the arrival of the goods.

Ⅲ. Closed Conversation

Directions: Please fill in the blanks according to the Chinese, then play roles with your partner.

(A: customer   B: secretary)

A: I'm satisfied with your product, but _____ (但是我想知道你们的售后服务怎么样?)

B: You may rest assured. _____ (我们关注产品的声誉,并且会尽力使顾客愉快的。)

A: If there's something wrong with the product, will it be replaced or repaired for free?

B: _____ (我们只在一个月内包换,) but we offer free repair within three years from the purchase date as long as you keep the receipt and the warranty card.

A: _____ (操作时遇到问题怎么办?)

B: You may contact us by phone, letter or E-mail. _____ (我们的专业人员会负责解决问题的。)

A: That's quite good. Thank you.

Ⅳ. Semi-Open Conversation

Directions: Please fill in the blanks with your own words, then play roles with your partner.

(A: customer   B: sales manager)

B: May I help you, Madam?

A: _____ .

B: What's the matter?

A: _____ .

B: Is that so? So far we haven't had any complaint of this kind.

A: I've brought some of them. _____ .

B: Oh, I'm terribly sorry. _____ .

A: How do you deal with the problem?

B: Please return the wrong pieces for our account. We'll deliver a replacement within two days.

A: _____ .

Ⅴ. Live Show

Challenge 1

Directions: A customer complained about the notebook he bought and asked to repair or replace it. You are in charge of dealing with the matter. Please act it out with your partner or partners. The model and the useful expressions below are just for your reference.

Challenge 2

Directions: You call your customers to collect their response to the products and solve the problems. Please act it out with your partner or partners. The model and the useful expressions below are just for your reference.

Model

(A: Lucy, salesgirl   B: Mr. Black, customer)

A: Hello, is that Mr. Black please?

B: Yes, who's that speaking?

A: This is Lucy speaking, from Customer Service Department of Joy Trade Company. I'm calling to ask if the delivery of your **projector** went OK.

B: Your delivery is very efficient. And the workers were very patient to tell me how to operate and maintain it.

A: It sounds good. Is there anything wrong with the projector or do you have any questions about it?

B: No, it is working very well. The **manual** is helpful to solve my problems.

A: You can contact us at all times. After receiving your trouble call, we'll visit you in 24 hours.

B: Thank you. By the way, what is the customer service hotline?

A: 800-12345678.

B: 800-12345678. I've got it. Can I update the product?

A: Yes. We'll update it for free within one year. And after that you can replace it with a new one at 30% discount.

B: Really? That's good news.

A: Mr. Black, please make comments on our service to help us improve our work.

B: Quite good. I'm very satisfied. I'll **recommend** your company to my colleagues and choose more products.

A: Thanks for your trust and encouragement. We won't **fail** you.

Notes

projector *n.* 投影仪   manual *n.* 说明书   recommend *v.* 推荐   fail *v.* 使失望

Useful expressions

1. Our customer satisfaction scores are among the best in the industry.
2. For any questions, please call our customer service hotline at 400-12345678.
3. We'll deliver and install it for you.
4. When do you want to deliver it?
5. Do you want the product repaired or replaced?
6. If there is any fault, you can replace it.

7. The factory offers a four-year warranty for this computer.
8. Please send it to our address for repair.
9. Because we're not careful about quality control, there are some faulty materials.
10. Can you tell me the brand and model?
11. Goods purchased in this plant can be changed in 7 days.

WRITING

E-mail

随着因特网的普及,电子邮件(E-mail)已成为最常用的高效、快速、成本低廉的通信方式,它不仅可以传递文字信息,还可以发送图片、音频、视频、动画文件等,是传统的通信方式无法比拟的。电子邮件主要由8个部分构成,包括 To:(收件人地址)、From:(寄件人地址)、Cc:(抄送)、Bcc:(暗抄送)、Subject:(主题)、Date:(日期)、Grade:(发送级别)和 Attachment:(附件)。并非每封邮件都需要这8个部分,但是收件人地址、寄件人地址、主题和日期是必不可少的。写电子邮件时要注意,语言简明直接,主题清晰具体,对收到的邮件及时查看、回复。

Sample 1

From: sally1978@sina.com
To: shanghai2010@yahoo.com
Date: April 7th, 2011, 8:05 a.m.
Subject: Purchase of Transmitters

James Smith,

I invite you to my office at 9:00 a.m. next Monday to discuss the purchase of some **transmitters** from your company. I hope through your visit we can settle the price and **conclude** the business before long. Please contact me soon and let me know whether it is convenient to you.

Regards,
Sally Lee

Notes

transmitter *n.* 发射机　　　conclude *v.* 结束

Sample 2

From: directorinwuhan@163.com
To: happylife@sohu.com

Cc: loveforever@craftmade.com
Date: May 17, 2011, 5:23 p.m.
Subject: **Negotiation** of the Prices
Grade: ordinary
Miss Zheng Lan,

I've read the catalogues and price lists in the attached E-mail you sent me, and I find your prices are much higher than your competitors'. It would be impossible for us to push any sales at such high prices. If you are prepared to cut down your prices by 8%, we might **come to terms**. I'm looking forward to your definite reply.

Yours sincerely,
Lisa Brown

Notes

negotiation n. 磋商　　　　　come to terms 达成协议

Challenge

Please write an E-mail to reply to Sample 2.

EXTENSIVE READING

5.5　Everyday Use of Lasers

In everyday life we're more or less surrounded by laser applications. Carpenters use lasers instead of spirit levels, hunters use lasers instead of ordinary telescopic sights and most likely, you use lasers when you listen to music.

DVD

A DVD player contains a laser that is used not because it produces a parallel beam, but rather because the light emerges from a tiny point, which enables it to be focused on the different layers of the disc. By moving the lens sideways—laterally, it is possible to reach areas farther in or out on the disc. By moving the lens along the beam—longitudinally, different depths can be reached in the disc. The information, ones and zeros, is stored in several layers, and only one layer is to be read at a time. Every point on a particular layer is read during every revolution of the disc.

In order to make room for a lot of information on every disc, the beam has to be focused on as small an area as possible. This cannot be done with any other light source than a laser.

Speed Measurement Using a Laser

The method the police use to measure car speed is based on a laser signal that is sent towards the target. This beam bounces back and is mixed with light that has not hit the car. The result is an oscillation—the same as when you tune a guitar—with higher frequency the faster the target moves.

The speed has to be measured straight from the front or from the back. If it is measured at an angle, the speed is underrated. This means that you cannot get false values that are too high.

The measurement is dependent on the car having something that reflects well. The license plate is perfect, as are different types of reflecting objects. Fogged surfaces are okay, but it will reduce the maximum distance.

Laser Distance Meter

The primary users of laser distance meters today are surveyors and constructors, but the car industry is catching on. Least spectacular is the so-called parking assistance that helps the driver to estimate the distance to the car behind when parking. A more recent application measures the distance to the car in front of the driver when driving on highways or other roads. You simply lock in the distance to the car in front of you in order to maintain that distance. This makes driving more efficient and faster as long as it all works. This kind of laser is found in most robots with mechanical vision.

Optical Loudspeaker Cable

Any amplifier of worth nowadays has an optical cable for transmission to the loudspeakers. The advantage of this method is that it is insensitive to interference from electromagnetic fields, which is interference from electronic devices and radio transmitters such as cell phones. The light source used as a transmitter is a small laser semiconductor. All equipment using optic cable uses the same standard. For example, the maximum bit rate for broadband applications is today 50-100 times higher using optics, but the potential ratio is 10,000 times.

Directions: Mark T (for True) if the statement agrees with the information given in the text; F (for False) if the statement contradicts the information given in the text.

(   )1. A DVD player contains a laser because the light emerges from a tiny point, which enables it to be focused on the different layers of the disc.

(   )2. In order to make room for a lot of information on every disc, the beam has to be focused on large space.

(   )3. When using a laser to measure the speed, the speed should be measured at an angle, not from the front or from the back.

(   )4. Parking assistance helps the driver to estimate the distance to the car behind when parking.

(   )5. Optical loudspeaker cable is insensitive to interference from electromagnetic fields.

5.6　Laser Safety

Laser Beam Hazards

The laser produces an intense, highly directional beam of light. If directed, reflected, or focused upon an object, laser light will be partially absorbed, raising the temperature of the surface and the interior of the object, potentially causing an alteration or deformation of the material.

Lasers have the potential to damage both the eye and the skin.

The human body is vulnerable to the output of certain lasers, and under certain circumstances, exposure can result in damage to the eye and skin. It is now widely accepted that the human eye is almost always more vulnerable to injury than human skin. The cornea, unlike the skin, does not have an external layer of dead cells to protect it from the environment. In the far-ultraviolet and far-infrared regions of the optical spectrum, the cornea absorbs the laser energy and may be damaged.

Non-Beam Hazards

In addition to the direct hazards to the eye and skin from the laser beam itself, it is also important to address other hazards associated with the use of lasers.

These non-beam hazards, in some cases, can be life-threatening, e.g. electrocution, fire, and asphyxiation. Because of the diversity of these hazards, the employment of safety and industrial hygiene personnel to effect the hazard evaluations may be necessary.

Safety Precautions Applicable to Solid-State Lasers

Enclosure of the beam and target in an opaque housing is the safest way of operating a laser. This level of safety precaution is **mandatory** for laser materials-processing systems operated in an industrial environment. In these systems interlocked doors, warning signs and lights, key-locked power switches, and emergency circuit breaker, and like precautions are taken to protect operators and passers-by from electrical and radiation hazards of the laser equipment. Also, microscopes and viewing ports are filtered or blocked to prevent the issuance of laser radiation, and laser impact points are surrounded by shields. At points of access for routine maintenance and set-up, warning signs are displayed prominently, and interlocks prevent firing of the laser while doors or ports are open.

In the laboratory it is often not possible to enclose fully a high-power laser. In these situations a series of safety precautions should be observed, e.g. wear laser safety glasses at all times! Never look directly into the laser light source or at scattered laser light from any reflective surface, etc.

Exercise

Directions: Choose the best answer from the four choices.

1. According to the first paragraph, laser light can _____ the surface of the object and alter the material.

    A. evaporate   B. cool off   C. break   D. heat up

2. Which of the following statement is not true? _____

    A. Lasers do damage to the eye rather than the skin.

    B. The cornea has an external layer of dead cells to protect it from the environment.

    C. The human body is easy to be destroyed by the output of lasers.

    D. The cornea absorbs the laser energy in the far-ultraviolet and far-infrared regions of the optical spectrum and may be damaged.

3. What is the meaning of the word "mandatory" in Para. 6?

    A. beneficial   B. effective   C. compulsory   D. imaginable

4. To prevent the issuance of laser radiation, safety personnel should _____.

    A. filter and block the microscopes and viewing ports

    B. show warning signs and light

    C. protect operators from radiation hazards

    D. prevent firing of the laser

5. In an industrial environment, the safest way of operating a laser is to _____.
A. wear laser safety glasses at all times
B. never look directly into the laser light source
C. enclose the beam and target in a non-transparent housing
D. take precautions to protect passer-bys from electrical hazards

TRANSLATION

How to Translate with Aid of the Internet and Software

翻译网站及软件在专业英语翻译中的应用

随着电脑及网络的普及,网络和电脑软件正成为人们进行英汉翻译时重要的帮手,但是不加甄别地使用网络及电脑软件,会犯下许多错误,甚至是严重错误,招来笑话。以下文为例。

A technique that has had recent success is laser cooling. This involves atom trapping, a method where a number of atoms are confined in a specially shaped arrangement of electric and magnetic fields. Shining particular wavelengths of laser light at the ions or atoms slow them down, thus cooling them. As this process is continued, they all are slowed and have the same energy level, forming an unusual arrangement of matter known as a Bose-Einstein condensate.

Google 翻译的译文如下。

最近成功的一个技术是激光冷却。这涉及原子捕集,原子在电场和磁场的特殊形状的安排只限于的方法。光辉的离子或原子在特定波长的激光灯减缓下来,从而冷却。由于这个过程是持续的,它们都放缓,有相同的能量水平,形成了一个不寻常的安排被称为玻色-爱因斯坦凝聚体的物质。

有道翻译的译文如下。

一种技术,是激光冷却最近成功。这涉及原子捕获方法,在那里有好些原子关在特殊形状的安排电场和磁场。闪亮的特定波长的雷射光,原子在离子或使得他们慢下来,因此冷却。这个过程是不断减少,他们都是有相同的能量水平,形成一个不寻常的事安排称为玻色－爱因斯坦凝聚态。

正确译文如下。

激光冷却技术最近获得成功。该技术涉及原子捕集,即将众多原子限制在特殊排列的电磁场里。将特定波长的激光束投射到离子或原子上，会使它们速度放慢，从而冷却它们。随着这一进程继续，它们的速度均放慢，并有相同的能级，形成了特别的物质组合，被称为玻色－爱因斯坦凝聚态。

综合比较三篇译文，可总结出网络翻译的一些特点。

（1）网络翻译在翻译专业术语或词汇方面具有一定参考性，如将 Bose-Einstein condensate 译为"玻色－爱因斯坦凝聚态"。

（2）网络翻译缺乏全局性，不可能站在全段甚至全篇的高度来把握句子和单词。网络翻译对句子的翻译是基于对单词和短语的对等翻译，并将其拼接而成的，因此，译文零碎、僵硬，如将 A technique that has had recent success is laser cooling 译为"一种技术，是激光冷却最近成功"。

（3）网络翻译具有机械性，过于局限于词与词的对等翻译，如在对 Shining particular wavelengths of laser light at the ions or atoms slows them down, thus cooling them 翻译时，出现了"光辉的离子或原子"这样令人咋舌的译文。

（4）网络翻译的正确性与句子的复杂度成反比，即句子越复杂，网络翻译的正确性越差。如 This involves atom trapping, a method where a number of atoms are confined in a specially shaped arrangement of electric and magnetic fields 的译文。

那么，正确应用网络及软件工具的翻译策略应如下。

（1）如上文所示，充分认识网络及软件工具在翻译时的利弊，用其长而避其短。

（2）掌握不同翻译网站及翻译软件的特色，有区别地综合运用，如有道翻译搜索功能强大，同一单词或词组的翻译，往往能提供多种选择。CNKI 翻译助手基于中国知网数据库，其专业词汇的翻译具有一定参考价值。

（3）翻译网站及翻译软件只能作为翻译的辅助工具，有时在翻译单词和短语时能提供一定参考，但一定要结合上下文，先甄别后采用。

（4）译文的准确性、流畅性、专业性主要取决于译者的中英文功底、翻译经验和专业知识的积累。

Try Out

Directions：*Please translate the following passages.*

1. Laser Printer

Producing near-professional quality print, the laser printer operates like a copying machine—with one important difference that the copying machine produces its image by focusing its lens on paper, while the laser printer traces an image by using a laser beam controlled by the computer. Laser printers produce text and graphics with high resolution; however, the print quality is not as high as that obtained by a phototype machine used by commercial textbook publishers. There are many inexpensive laser printers that print in black and white on the market today, and it is a good choice because of its beautiful graphics capabilities.

2. 九头鸟激光集团简介

1990年成立之初，九头鸟激光集团仅生产焊接激光机。经过20多年的发展，公司生产的各种类型激光机工业用途广泛,如激光打标、激光打孔、激光切割、激光热处理,并与欧美最新设计和技术同步升级。秉承"服务顾客、创造价值"的理念,公司的核心竞争力不断得以提升。截至目前,公司在海内外共售出 7 000 多套工业激光机。

TASK

Laser Marking

Target	Understand and express professional knowledge.				
Requirements	Make an English presentation according to lesson 5.2. Produce a micro-video to express professional knowledge. Add accurate Chinese and English subtitles to the video.				
Contents	Laser Marking: Hot working and cool working Advantages The physical principle of laser marking Three categories				
Group: _____	Project	Name	Software	Time	Score
	Material collection				
	Script				
	PPT				
	Reader				
	Subtitle				
	MP4				
Operating environment	XP/Win7/Win8/ Mobile phone				
Product features					

Script reference

标题	专业内容（讲稿与中、英文字幕）	PPT 中需显示的文字	表现形式	素　材
每个知识点的要点或者小节名字	做报告时的英文讲稿视频中的中、英文字幕	需要特别提醒的文字	讲解到某个地方或某句话时，出现的图片、动画或者文字	照片或者视频
Laser Marking 激光打标	Laser marking is a non-contact thermal process that uses the heat generated by a laser beam to alter the surface of the workpiece. 激光打标是一个非接触性的热反应过程，通过激光束所产生的热能来改变工件的表面。	Laser marking is a non-contact thermal process that uses the heat generated by a laser beam to alter the surface of the workpiece.	激光打标效果图	图1 激光打标，效果图
	Contents Hot working and cool working Advantages compared with other marking technologies The physical principle of laser marking The different markings can be classified in three categories 主要内容： 热加工和冷加工 与其他的打标技术比较 激光打标的物理原理 打标可以分为三类	Contents 1. Hot working and cool working 2. Advantages 3. The physical principle of laser marking 4. Three categories	文字显示微课主要内容	

续表

标题	专业内容(讲稿与中、英文字幕)	PPT中需显示的文字	表现形式	素 材
1. Hot working and cool working 热加工和冷加工	At the present, there are two types recognized principles: by hot working and by cool working. 目前有两种公认的打标原理:热加工和冷加工。	By hot working and by cool working.		
Hot working 热加工	Hot working uses laser beam with higher energy density to irradiate on the materials. The materials become hotter by absorbing the laser energy and with phenomena of melting, ablation and evaporation, etc. 热加工使用高能量密度的激光束照射在材料表面,材料通过吸收激光能量变热,从而发生熔化、切除和蒸发等现象。	Hot working uses laser beam with higher energy density to irradiate on the materials. The materials become hotter by absorbing the laser energy and with phenomena of melting, ablation and evaporation, etc.	文字	
Cool working 冷加工	Cool working uses high energy ultraviolet photons to break materials (especially organic materials) or chemical bonds in the mediums around, to destroy materials in the non-hot process. 冷加工使用高能量的紫外光破坏材料(尤其是有机材料)或在介质中的化学键,从而破坏无热加工中的材料。	Cool working uses high energy ultraviolet photons to break materials (especially organic materials) or chemical bonds in the mediums around, to destroy materials in the non-hot process.	文字	
	Cool working processing has special significance for laser marking, because it does not work through hot working, and will not produce hot damage or break cold stripping of chemical bonds, therefore, no hot deformation will be caused on the surface of the materials. 冷加工对于激光打标具有特别的意义,因为它不通过热加工,不会产生热损伤,也不会打破化学键的冷剥离,因此,材料表面不会变形。			

续表

标题	专业内容（讲稿与中、英文字幕）	PPT 中需显示的文字	表现形式	素　材
2. Advantages 优点	Compared with other marking technologies such as ink jet printing and mechanical marking, laser marking has a number of advantages, 与其他的打标技术如喷墨打印和机械打标等相比，激光打标有很多优势，			
	such as very high processing speeds, low operation cost (no use of consumables), constant high quality and durability of the results, avoiding contaminations, the ability to write very small features, and very high flexibility in automation. 比如高加工速度，低操作成本（不需要使用耗材），稳定的高质量以及结果耐久，同时避免了污染，还能够标记很小的特征，在自动操作中也有较高的灵活性。	High processing speeds Low operation cost (no use of consumables) Constant high quality Durability of the results Avoiding contaminations Ability to write very small features High flexibility in automation		
3. The physical principle of laser marking 激光打标的物理原理	The physical principle of laser marking is based on light amplification by simulated emission of radiation. 激光打标的物理原理是基于"通过辐射受激发射的光放大"。	The physical principle of laser marking is based on light amplification by simulated emission of radiation.		图 2 激光打标原理图
	In other words, the energy emitted by a source (e.g. arc lamp) onto an active source (e.g. Nd：Yag crystal) will be collected and concentrated between two opposing mirrors. 换言之，由源极（如弧光灯）发出照到放射源（如掺钕的钇铌石榴石晶体）之上的能量将被集中在两面对立的镜子中。		对照原理图讲解	图 2 激光打标原理图

续表

标题	专业内容(讲稿与中、英文字幕)	PPT中需显示的文字	表现形式	素材
3. The physical principle of laser marking 激光打标的物理原理	**Part 1** The oscillation of specific light particles (photons) that move between the mirrors generates a high power laser beam. Only part of this beam passes through the output mirror (typically 10 to 20%), the rest being reinjected for the amplification process: the laser beam is thus created. 在镜子间移动的特殊光微粒(光子)的振动产生了高能量的激光束。这一光束,一部分穿过输出镜(通常是10%到20%),剩余的光束再次射入参与放大光束的过程,激光束因而产生了。	**Part 1** The oscillation of specific light particles (photons) that move between the mirrors generates a high power laser beam. Only part of this beam passes through the output mirror (typically 10 to 20%), the rest being reinjected for the amplification process: the laser beam is thus created.		图2 激光打标原理图
	Part 2 The beam is deflected by two scanning mirrors, driven electronically and via software, thus enabling a very high scanning speed and exceptional positioning accuracy. 两面振镜使激光束偏转,通过电驱动和软件,可以产生很高的扫描速度和惊人的定位准确性。	**Part 2** The beam is deflected by two scanning mirrors, driven electronically and via software, thus enabling a very high scanning speed and exceptional positioning accuracy.		图2 激光打标原理图
	Part 3 The flat field lens then concentrates the beam energy in a very tiny spot and greatly increases the power density on the surface to be marked. 场镜在很小的点上聚集光束能量,并在需要打标的材料表面增加能量密度。	**Part 3** The flat field lens then concentrates the beam energy in a very tiny spot and greatly increases the power density on the surface to be marked.		图2 激光打标原理图

续表

标题	专业内容（讲稿与中、英文字幕）	PPT中需显示的文字	表现形式	素　材
4. Three categories 三类	The different markings can be classified in three categories： 打标可以分为三类： Engraving with material removal (sublimation) Material annealing with surface color change Marking through layer removal	The different markings can be classified in three categories： Engraving with material removal (sublimation) Material annealing with surface color change Marking through layer removal		图3 通过材料切除来雕刻（升华） 图4 通过退火使材料表面颜色发生改变 图5 通过切除涂层进行打标
Engraving with material removal (sublimation) 通过材料切除来雕刻（升华）	The energy is delivered with high peak power pulses so that the material is instantaneously removed without thermal side-effects on the parts. 能量由高峰值能量脉冲发出，因此材料可以瞬间被移除，而不会在部件上产生热负效应。	Engraving with material removal (sublimation)		图3 通过材料切除来雕刻（升华） material surface ablation
Material annealing with surface color change 通过退火使材料表面颜色发生改变	The energy is delivered with lower pulses, heating the material and changing the surface appearance. No material is removed. 低脉冲产生能量，加热材料并改变表面外形。无须切除材料。	Material annealing with surface color change		图4 通过退火使材料表面颜色发生改变 surface color change
Marking through layer removal 通过切除涂层进行打标	On coated material, the contrast is created by removing the top layer, thus showing the color of the base material. 在有涂层的材料上，通过切除涂层顶层造成对比，因而显示出基底材料的颜色。	Marking through layer removal		图5 通过切除涂层进行打标 one layer removed two layers removed

UNIQUE ENGLISH

Funny Proverbs

1. Behind every successful man, there is a woman. And behind every unsuccessful man, there are two.
2. Success is a relative term. It brings so many relatives.
3. Love is like a photo. It needs darkness to develop.
4. Your future depends on your dreams. So go to sleep.
5. "Hard work never killed anybody." But why take the risk?
6. God made relatives. Thank God we can choose our friends.

Unit 6
Fiber-Optic Communication

In this unit, you are going to learn the following contents:
READING IN DEPTH
6.1　Brief Introduction of Fiber-Optic Communication
6.2　Typical Optical Transmitter—LEDs
6.3　Typical Optical Receivers
6.4　Comparison with Electrical Transmission
CAREER SPEAKING
Business Meeting
WRITING
Memo
EXTENSIVE READING
6.5　Optical Fibers
6.6　Fiber to the Home
TRANSLATION
Translation of Adverbial Clauses
TASK
Fiber-Optic Communication
UNIQUE ENGLISH
Mother

READING IN DEPTH

6.1 Brief Introduction of Fiber-Optic Communication

Learning objectives:

In this section, you will:
- understand the meaning of fiber-optic communication.
- learn about the development of optical fiber communication.
- comprehend how the fiber-optic communication works.
- learn the advantages and obstacles of the optical fiber compared with the copper wire.

Warm-Up Activity

The following pictures show us common communication devices in our daily life. Please write down what they transmit. You may work with your partner.

Cables

Picture 1

Radio

Picture 2

Beacon Tower
Picture 3

Fibers
Picture 4

Text

What is fiber-optic communication? It is a method of transmitting information from one place to another by sending pulses of light through an optical fiber.

Optical fiber was successfully developed in 1970 by Corning Glass Works, with **attenuation** low enough for communication purposes (about 20 dB/km), and at the same time GaAs **semiconductor** lasers were developed that were compact and therefore suitable for transmitting light through fiber optic cables for long distances.

After a period of research starting from 1975, the first commercial fiber-optic communication system was developed, which operated at a wavelength around 0.8 μm and used GaAs semiconductor lasers. This first-generation system operated at a bit rate of 45 Mbps with **repeater** spacing of up to 10 km.

The light forms an electromagnetic carrier wave that is **modulated** to carry information. First developed in the 1970s, fiber-optic communication systems have revolutionized the telecommunications industry and have played a major role in the **advent** of the information age. Because of its advantages over electrical transmission, optical fibers have largely replaced copper wire communications in core networks in the developed world.

attenuation *n.* 衰减,变薄,稀释

semiconductor *n.* 半导体

repeater *n.* 中继器；转发器

modulate *vt.* （信号）调制

advent *n.* 出现,到来

And how exactly does the fiber-optic communication work? Simply speaking, there are 3 steps in short.

(1) Use an **optical transmitter** to create the optical signal.

(2) Relay the signal along the fiber, make sure the signal does not become too **distorted** or weak.

(3) Use a **receiver** to receive the optical signal, and **converting** it into an electrical signal.

Optical fiber is used by many telecommunications companies to transmit telephone signals, Internet communication, and **cable television** signals. Because much lower attenuation and **interference**, optical fiber has large advantages over existing copper wire in long-distance and high-demand applications. However, **infrastructure** development within cities was relatively difficult and time-consuming, and fiber-optic systems were complex and expensive to install and operate. Due to these difficulties, fiber-optic communication systems have primarily been installed in long-distance applications, where they can be used to their full transmission capacity, **offsetting** the increased cost. Since 2000, the prices for fiber-optic communications have dropped considerably. The price for rolling out fiber to the home has currently become more **cost-effective** than that of rolling out a copper based network.

Since 1990, when optical-**amplification** systems became commercially available, the telecommunications industry has laid a vast network of intercity and transoceanic fiber communication lines. By 2002, an **intercontinental** network of 250 000 km of submarine communications cable with a capacity of 2.56 Tb/s was completed, and although specific network capacities are **privileged** information, telecommunications investment reports indicate that network capacity has increased dramatically since 2004.

intercontinental *adj.* 洲际的，大陆间的
privileged *adj.* 特许的，专用的

Get to Work

Ⅰ. Matching

Directions: Match the words or expressions in the left column with the Chinese equivalents in the right column.

1. optical fiber
2. optical-amplification
3. interference
4. attenuation
5. modulate
6. distort
7. cable
8. transmitter
9. receiver
10. electromagnetic

A. 光纤
B. 调制
C. 电缆
D. 电磁的
E. 发射器
F. 使失真
G. 接收器
H. 衰减
I. 干扰
J. 光放大器

Ⅱ. Understanding Checking

Directions: Mark Y (for YES) if the statement agrees with the information given in the text; N (for NO) if the statement contradicts the information given in the text; NG (for NOT GIVEN) if the information is not given in the text.

( )1. Fiber-optic communication system is barely used in Internet communication.
( )2. Fiber-optic communication system includes transmitters and receivers.
( )3. Too expensive to install and operate is the reason that fiber-optic communication systems have been installed in long-distance applications.
( )4. Fiber-optic communication systems were first developed in the 1970s.
( )5. Lasers are transmitted as the optical signal.

Ⅲ. Passage Completion

Directions: Fill in each of the blanks with one of the words or expressions in the box, making changes if necessary.

formed	operating	became	took over
went into	led to	merged	in

Advanced Fibreoptic Engineering (AFE) was __1__ in June 2005.

However, a fibreoptic business has been __2__ in our Witney premises since 1994. Some customers may even remember the same business running from "The Ropery" in Burford before that.

The initial business in Burford traded as GCA Electronics and distributed electronic components. One line of products, optical fibre coupling LEDs for ABB Hafo __3__ Sweden, __4__ a demand for fibre alignment. This manufacturing element soon dominated the distribution business and distribution was discontinued. The company __5__ GCA Fibreoptics.

In 1998 Uniphase __6__ with JDS Fitel to form JDS Uniphase (JDSU).

During 2001 a management buy out (MBO) was formed and Afonics Fibreoptics __7__ the Witney facilities, Yarnton being far too large. Afonics grew from 2001 through to 2005, but never achieved profitability. In May 2005 Afonics Fibreoptics __8__ administration.

Ⅳ. **Translation**

Directions: Complete the sentences by translating into English according to the Chinese given in brackets.

1. Relay the signal along the fiber, _____（确保信号不会失真或衰减）.

2. _____（因为相较很低的失真的干扰）, optical fiber has large advantages over existing copper wire in long-distance and high-demand applications.

3. Optical fiber is used by _____（电信公司传输电话信号、互联网通信和有线电视信号）.

4. _____（铺设光纤的价格目前已经越来越具有更高的性价比）than that of rolling out a copper based network.

5. By 2002, an intercontinental network of 250 000 km of _____ （传输速度为 2.56 Tb/s 的海底通信电缆）was completed.

Ⅴ. **Sentence Remaking**

Directions: Simulate the following sentence patterns according to an example provided from the text.

1. ... have revolutionized ... and have played a major role in ...

E. g. Fiber-optic communication systems have revolutionized the telecommunications industry and have played a major role in the advent of the information age.

_____ have revolutionized _____ and have played a major role in _____ .

2. ... is used by ... to ...

E. g. Optical fiber is used by many telecommunications companies to transmit telephone signals.

_____ is used by _____ to _____ .

3. Due to…, … has large advantages over…

E. g. Due to much lower attenuation and interference, optical fiber has large advantages over existing copper wire in long-distance and high-demand applications.

Due to _____, _____ has large advantages over _____.

4. … been installed in… where….

E. g. Fiber-optic communication systems have primarily been installed in long-distance applications, where they can be used to their full transmission capacity, offsetting the increased cost.

_____ been installed in _____ where _____.

5. Because…, … has large advantages over…

E. g. Because much lower attenuation and interference, optical fiber has large advantages over existing copper wire in long-distance and high-demand applications.

Because _____, _____ has large advantages over _____.

6.2 Typical Optical Transmitter—LEDs

Learning objectives:

In this section, you will:
- learn about the difference between LEDs and laser diodes.
- comprehend the operational principle of LEDs.
- master the advantages of LEDs.
- find the applications of LEDs.

Warm-Up Activity

Can you identify the positive pole and the negative pole of this device? Do you know how to install it at the circuit to make the device give out light? Please draw it on your book.

Text

The most commonly-used optical transmitters are semiconductor devices such as light-emitting diodes (LEDs) and **laser diodes** (LD). The difference between LEDs and laser diodes is that LEDs produce **incoherent light**, while laser diodes produce **coherent light**. For use in optical communications, semiconductor optical transmitters must be designed to be compact, efficient, and reliable, while operating in an optimal wavelength range, and directly modulated at high **frequencies**.

In its simplest form, an LED is a forward-biased **p-n junction**, emitting light through spontaneous emission, a **phenomenon** referred to as **electroluminescence**. The emitted light is incoherent with a relatively wide spectral width of 30—60 nm. Communications LEDs are most commonly made from GaAsP or GaAs. Because GaAsP LEDs operate at a longer wavelength than GaAs LEDs (1.3 mm vs. 0.81—0.87 mm). The large spectrum width of LEDs causes higher fiber **dispersion**, considerably limiting their bit rate-distance product (a common measure of usefulness). LEDs are suitable primarily for local-area-network applications with bit rates of 10-100 Mb/s and transmission distances of a few kilometers.

laser diode
激光二极管
incoherent light
非相干光
coherent light
相干光
frequency n. 频率, 频繁
p-n junction
p-n 结

phenomenon n. 现象
electroluminescence n.
[电子] 场致发光, 电致发光

dispersion n. 散布

The low energy **consumption**, low **maintenance** and small size of modern LEDs has led to uses as **status indicators** and displays on a variety of equipment and installations. Large-area LED displays are used as **stadium** displays and as **decorative** displays. Thin, lightweight message displays are used at airports and railway stations, and as destination displays for trains, buses, **trams**, and **ferries**.

With the development of high efficiency and high power LEDs it has become possible to use LEDs in lighting and **illumination**. Replacement light **bulbs** have been made. LEDs are used as street lights and in other **architectural** lighting where color changing is used. The **mechanical robustness** and long lifetime is used in automotive lighting on cars, motorcycles and on bicycle lights. LED street

lights are used on poles and in parking garages. In 2007, the Italian village Torraca was the first place to convert its entire illumination system to LEDs.

LEDs are used in **aviation** lighting. Airbus has used LED lighting in their Airbus A320 Enhanced since 2007, and Boeing plans its use in the 787. LEDs are also being used now in airport and **heliport** lighting. LED airport **fixtures** currently include medium-intensity runway lights, runway centerline lights, **taxiway** centerline & edge lights, guidance signs and obstruction lighting. LEDs are also suitable for backlighting for LCD televisions and lightweight **laptop** displays and light source for DLP projectors (see LED TV). RGB LEDs raise the color **gamut** by as much as 45%. Screens for TV and computer displays can be made thinner using LEDs for backlighting.

consumption n. 消费，消耗
maintenance n. 维护，维修
status indicator 状态指示器
stadium n. 体育场，露天大型运动场
decorative adj. 装饰性的，装潢用的
tram n. 有轨电车
ferry n. 渡轮，摆渡
illumination n. 照明
bulb n. 电灯泡
architectural adj. 建筑学的，建筑上的
mechanical adj. 机械的，力学的
robustness n. 稳健性，健壮性
aviation n. 航空
heliport n. 直升机机场
fixture n. 设备，固定装置
taxiway n. 滑行道

laptop n. 便携式电脑，笔记本电脑
gamut n. 整个范围

Get to Work

Ⅰ. Matching

Directions: Match the words or expressions in the left column with the Chinese equivalents in the right column.

1. semiconductor   A. 现象
2. light-emitting diode   B. 能量损耗
3. laser diode   C. 笔记本电脑
4. incoherent light   D. 照明
5. frequency   E. 场致发光
6. electroluminescence   F. 非相干光
7. illumination   G. 发光二极管
8. laptop   H. 半导体
9. energy consumption   I. 频率
10. phenomenon   J. 激光二极管

Ⅱ. Understanding Checking

Directions: Mark Y (for YES) if the statement agrees with the information given in the text; N (for NO) if the statement contradicts the information given in the text; NG (for NOT GIVEN) if the information is not given in the text.

(　　)1. According to the text, LED is a kind of metal device.
(　　)2. GaAsP LEDs operate at a wavelength of 0.81—0.87 mm.
(　　)3. LEDs are suitable for short distance transmission.
(　　)4. Nowadays, many families have LED TV sets.
(　　)5. Due to LEDs' different kinds of colors, they are used in lighting and illumination.

Ⅲ. Passage Completion

Directions: Fill in each of the blanks with one of the words or expressions in the box, making changes if necessary.

over	comprised	bit	instead	from
over	since	weighed	similar	onto

LEDs have come a long way ___1___ the early days of lighting up digital clock faces. In the 2000s, LCD TVs took ___2___ the high definition market and represented a huge step over old standard definition CRT televisions. LCD displays were even a major step ___3___ HD rear-projection sets that ___4___ well over 100 pounds (45.4 kilos). Now LEDs are poised to make a ___5___ jump. While LCDs are far thinner and lighter than massive rear-projection sets, they still use cold cathode fluorescent tubes to over a white light ___6___ the pixels that make up the screen. Those add weight and thickness to the television set. LEDs solve both problems.

Have you ever seen a gigantic flatscreen TV barely an inch thick? If you have, you've seen an LED television. Here's where the acronyms get a ___7___ confusing: those LED TVs are still LCD TVs, because the screens themselves are __8__ of liquid crystals. Technically, they're LED-backlit LCD TVs. __9__ of fluorescent tubes, LEDs shine light __10__ behind the screen, illuminating the pixels to create an image.

IV. Translation

Directions: Complete the sentences by translating into English according to the Chinese given in brackets.

1. _____ (半导体光发射器,例如发光二极管和激光二极管) is the most common optical transmitter in our daily life.

2. LED is widely used in TV because _____ (红绿蓝颜色表示法将颜色范围提高了 45%).

3. The light given off by LED _____ (是光谱范围为 30~60 nm 的不相干光).

4. With the development of high efficiency and high power LEDs it _____ (有可能将发光二极管应用在照明上).

5. Maybe someday in the future, we could _____ (将家庭照明系统换成 LED).

V. Sentence Remaking

Directions: Simulate the following sentence patterns according to an example provided from the text.

1. ...in the advent of...

E.g. Fiber-optical communication systems have played a major role in the advent of the information age.

_____ in the advent of _____ .

2. The difference between...is that...

E.g. The difference between LEDs and laser diodes is that LEDs produce incoherent light, while laser diodes produce coherent light.

The difference between _____ is that _____ .

3. ...has led to...

E.g. The low energy consumption, low maintenance and small size of modern LEDs has led to uses as status indicators and displays.

_____ has led to _____ .

4. With the development of...it has become possible to...

E.g. With the development of high efficiency and high power LEDs it has become possible to use LEDs in lighting and illumination.

With the development of _____ it has become possible to _____ .

5. ...was the first to...

E.g. The Italian village Torraca was the first place to convert its entire illumination system to LEDs.

_____ was the first to _____ .

6.3 Typical Optical Receivers

Learning objectives:

In this section, you will:
- learn about the main components of an optical receiver.
- compare several types of photodiodes and find their applications.

Warm-Up Activity

Which of the following devices contain optical receivers? Discuss with your partner and orally explain your reasons.

TV Set

Picture 1

IC Card Analyzer

Picture 2

Bluetooth Headset
Picture 3

Barcode Scanner
Picture 4

Text

As we all know, light can be detected by eyes. But the eye is not suitable for modern fiber communications because its response is too slow, its sensitivity to low-level signals is **inadequate**, and it is not easily connected to electronic receivers for amplification, **decoding**, or other **signal processing**. Furthermore, the spectral response of the eye is limited to wavelength between 0.4 μm and 0.7 μm, where fibers have high loss.

The main component of an optical receiver is a **photodetector**, which converts light into electricity using the **photoelectric effect**. The photodetector is typically a semiconductor-based **photodiode**. Several types of photodiodes include p-n photodiodes, a p-i-n photodiodes, and **avalanche photodiodes**.

PIN diodes are much faster and more sensitive than ordinary p-n junction diodes, and **hence** are often used for optical communications and in lighting regulation.

P-N photodiodes are not used to measure extremely low light intensities. Instead, if high sensitivity is needed, avalanche photodiodes, intensified **charge-coupled devices** or **photomultiplier tubes** are used for applications such as **astronomy**, **spectroscopy**, night vision equipment and laser rangefinding.

The PIN device is cheaper, is less sensitive to temperature, and requires lower **reverse bias** voltage than the APD. The speeds of the two devices are comparable, so the PIN diode is preferable in most systems. The APD gain is needed when the system is less limited, as occurs for long-distance links. Metal-semiconductor-metal (MSM)

inadequate *adj.* 不充分的，不适当的
decoding *n.* [通信] 解码
signal processing 信号处理

photodetector *n.* 光电探测器
photoelectric effect 光电效应
photodiode *n.* [电子] 光电二极管
avalanche photodiode 雪崩光电二极管
hence *adv.* 因此，今后
charge-coupled device 电荷耦合器
photomultiplier tube 光电倍增管
astronomy *n.* 天文学
spectroscopy *n.* [光] 光谱学

reverse bias 反向偏压

photo detectors are also used due to their suitability for **circuit integration** in wavelength-division multiplexers.

Photodiodes are used in consumer electronics devices such as compact disc players, smoke detectors, and the receivers for **remote controls** in VCRs and televisions.

In other consumer items such as camera light meters, clock radios (the ones that **dim** the display when it's dark) and street lights, **photoconductors** are often used rather than photodiodes, although in principle either could be used.

Photodiodes are often used for **accurate** measurement of light intensity in science and industry. They generally have a better, more linear response than photoconductors.

They are also widely used in various medical applications, such as detectors for **computed tomography** or instruments to analyze samples (**immunoassay**).

circuit *n.* 电路,回路
integration *n.* 集成,综合
remote control 遥控器
dim *vt.* 使暗淡,使失去光泽
photoconductor *n.* [物] 光电导体,光敏电阻

accurate *adj.* 精确的

computed tomography 计算机断层扫描
immunoassay *n.* 免疫分析,免疫测定

Get to Work

Ⅰ. Matching

Directions: Match the words or expressions in the left column with the Chinese equivalents in the right column.

1. decoding
2. photodetector
3. photoelectric effect
4. photodiode
5. avalanche photodiode
6. photoconductor
7. integration
8. consumer electronics
9. spectral response
10. circuit

A. 光电二极管
B. 集成
C. 回路
D. 家用电器
E. 光谱响应
F. 光敏电阻
G. 光电效应
H. 雪崩光电二极管
I. 解码
J. 光电探测器

II. Understanding Checking

Directions: Mark Y (for YES) if the statement agrees with the information given in the text; N (for NO) if the statement contradicts the information given in the text; NG (for NOT GIVEN) if the information is not given in the text.

(　　)1. The eye is not suitable for modern fiber communication because it is not sensitive enough.

(　　)2. The photodetector converts light into electricity.

(　　)3. All the photodetectors are semiconductor-based photodiodes.

(　　)4. The PIN device is preferable in most systems because its speed is much faster than APD.

(　　)5. Optical-electrical converters produce an analog signal in the electrical domain.

III. Passage Completion

Directions: Fill in each of the blanks with one of the words or expressions in the box, making changes if necessary.

| offering | tested | in | on | advice |
| instructions | access | savings | range | over |

Qphotonics is a specialized distributor of semiconductor light emitting devices __1__ 660—1800 nm wavelength __2__ for your research, development, and production. Each laser device is fully __3__ in house and characterized for performance. We focus __4__ establishing long-term relations with our customers by __5__ value in our products and service.

We offer efficient and inexpensive solutions to engineers and researchers looking for semiconductor light sources.

a. Large variety (__6__ 100 models).

b. Rapid __7__ to the right device, without compromising specifications or quality.

c. Detailed datasheets online.

d. Fast delivery (most laser diodes are in stock).

e. Price __8__ on standard products.

f. Unusual wavelengths, custom configurations fabricated in small lots.

g. __9__ on laser diode selection and operation.

h. Complementary products (laser controllers, mounts, photodiodes).

i. Simple operating __10__ and handouts.

j. Efficient shopping experience.

We strive for customer satisfaction by providing fully tested quality products, comprehensive datasheets, and operating manuals.

IV. Translation

Directions: Complete the sentences by translating into English according to the Chinese given in brackets.

1. Consumer electronics devices such as compact disc players, smoke detectors _____ _____ (都使用了光电二极管).

2. The eyes' response is _____ (局限在 0.4~0.7 μm 的波段内).

3. PIN diodes often used for optical communications and in lighting regulation _____ _____ (因为它比普通的 PN 结器件速度更快且更灵敏).

4. The PIN device is cheaper, is less sensitive to temperature _____ _____ (并且与雪崩二极管相比只要较低的反向电压).

5. (由于光电二极管较光敏电阻线性好、响应快) _____, photodiodes are often used for accurate measurement of light intensity in science and industry.

Ⅴ. Sentence Remaking

Directions: Simulate the following sentence patterns according to an example provided from the text.

1. ...is not suitable for... because...

E.g. The eye is not suitable for modern fiber communications because its response is too slow.

_____ is not suitable for _____ because _____ .

2. ...than..., and hence...

E.g. PIN diodes are much faster and more sensitive than ordinary p-n junction diodes, and hence are often used for optical communications and in lighting regulation.

_____ than _____ , and hence _____ .

3. ...are comparable,...is preferable in...

E.g. The speeds of the two devices are comparable, so the PIN diode is preferable in most systems.

_____ are comparable, _____ is preferable in _____ .

4. ...are used in...such as...

E.g. Photodiodes are used in consumer electronics devices such as compact disc players, smoke detectors.

_____ are used in _____ such as _____ .

5. ...in any condition

E.g. We cannot mix up the two kinds of devices in any condition.

_____ in any condition.

6.4 Comparison with Electrical Transmission

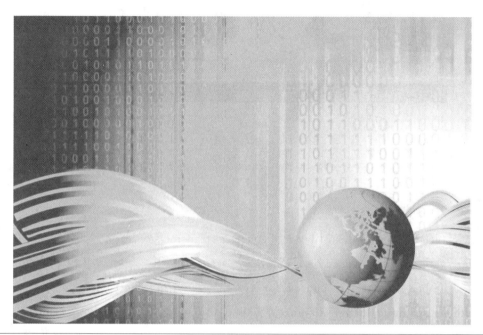

Learning objectives:
In this section, you will:
● learn about the main benefits of fiber.
● comprehend why in short distance and relatively low bandwidth applications, electrical transmission is often preferred.
● find the important features of fiber when fiber may be used even for short distance or low bandwidth applications.
● learn the installation of optical fiber cables.

Warm-Up Activity

Fill in the brackets with the words or expressions given below according to the wavelength of each electromagnetic wave.

X-ray	X 射线
infrared	红外线
microwave	微波
gamma ray	伽马射线
ultraviolet ray	紫外线
wireless	无线电
visible light	可见光

Text

The choice between optical fiber and electrical transmission for a particular system is made based on a number of **trade-offs**.

Optical fiber is generally chosen for systems requiring higher **bandwidth** or spanning longer distances than electrical cabling can accommodate.

The main benefits of fiber are its exceptionally low loss and its **inherently** high data-carrying capacity. Thousands of electrical links would be required to replace a single high bandwidth fiber cable. Another benefit of fibers is that even when run alongside each other for long distances, fiber cables experience effectively no **crosstalk**, in contrast to some types of electrical transmission lines. Fiber can be installed in areas with high electromagnetic interference (**EMI**), such as alongside **utility** lines, power lines, and railroad tracks.

In short distance and **relatively** low bandwidth applications, electrical transmission is often preferred because of its:

(1) lower material cost, where large quantities are not required;

(2) lower cost of transmitters and receivers;

(3) capability to carry **electrical power** as well as signals (in specially-designed cables);

(4) ease of operating **transducers** in linear mode.

trade-off 权衡

bandwidth n. [电子][物] 带宽

inherently adv. 内在地，固有地

crosstalk n. 串话干扰，串台
EMI abbr. 电磁干扰
utility n. 公共设施
relatively adv. 相当地，相对地

electrical power 电功率
transducer n. 传感器

In certain situations fiber may be used even for short distance or low bandwidth applications, due to other important **features**.

(1) **Immunity** to electromagnetic interference, including nuclear **electromagnetic pulses** (although fiber can be damaged by alpha and beta radiation).

(2) High electrical **resistance**, making it safe to use near high-voltage equipment or between areas with different earth **potentials**.

(3) Lighter weight—important, for example, in aircraft.

(4) No sparks—important in flammable or explosive gas environments.

(5) Not electromagnetically radiating, and difficult to tap without disrupting the signal—important in high-security environments.

(6) Much smaller cable size—important where pathway is limited, such as networking an existing building.

Optical fiber cables can be installed in buildings with the same equipment that is used to install copper and **coaxial cables**, with some **modifications** due to the small size and limited pull tension and bend radius of optical cables. Optical cables can typically be installed in duct systems in **spans** of 6 000 m or more depending on the duct's condition, layout of the duct system, and installation technique. Longer cables can be **coiled** at an intermediate point and pulled farther into the duct system as necessary.

Get to Work

Ⅰ. Matching

Directions: *Match the words or expressions in the left column with the Chinese equivalents in the right column.*

1. trade-off   A. 电功率
2. bandwidth   B. 免于, 抗……性
3. crosstalk   C. 电阻
4. coaxial cable   D. 公共设施
5. potential   E. 特点
6. feature   F. 权衡
7. immunity to   G. 同轴电缆
8. electrical power   H. 带宽
9. resistance   I. 潜在的
10. utility   J. 串台

Ⅱ. Understanding Checking

Directions: *Mark Y (for YES) if the statement agrees with the information given in the text; N (for NO) if the statement contradicts the information given in the text; NG (for NOT GIVEN) if the information is not given in the text.*

(   )1. Fiber-optic communication's best benefits are its exceptionally low loss and its inherently high data-carrying capacity.

(   )2. Because the fiber cables experience effectively no crosstalk, we can use them in the confidentiality communication.

(   )3. To lay a network system in an existing building, fiber cables are the better choice.

(   )4. Electrical transmission is often preferred in long distance transmit.

(   )5. Fiber-optic communication has taken the place of electrical transmission now.

Ⅲ. Passage Completion

Directions: *Fill in each of the blanks with one of the words or expressions in the box, making changes if necessary.*

deliver	refer	individual	during	looking
current	main	within	drive	increase

Stop and think how your Internet usage has evolved __1__ the last few years. If you're like most people, you're doing—and expecting—a lot more of your Internet like increased interactivity, rich media and uploading and downloading pictures and video.

More large files are moving across the cyberspace network these days, and experts expect that trend will only __2__ . A January 2008 study by the Discovery Institute estimates new technologies will __3__ Internet traffic up by 50 times its __4__ rate __5__ the next 10 years.

The pressure for better connectivity is one of the __6__ reasons providers and users are __7__ at fiber-to-the-home broadband connections as a potential solution.

Fiber-to-the-home broadband connections, or FTTH broadband connections, __8__ to fiber optic cable connections for __9__ residences. Such optics-based systems can __10__ a multitude of digital information—telephone, video, data, et cetera—more efficiently than traditional copper coaxial cable for about the same price. FTTH premises depend on both active and passive optical networks to function.

Ⅳ. Translation

Directions: Complete the sentences by translating into English according to the Chinese given in brackets.

1. _____（消耗较少的发射器和接收器）is one of the reason that electrical transmission is often preferred in short distance applications.

2. _____（一根光纤携带的信号量）equal to thousands of electrical links.

3. Fibers are preferred where pathway is limited _____（因为它的尺寸较小）.

4. If it is necessary _____（我们能将光纤卷曲起来）.

5. Lighter weight, _____（使得光纤较铜线便于运输）.

Ⅴ. Sentence Remaking

Directions: Simulate the following sentence patterns according to an example provided from the text.

1. The main benefits of ... Another benefit of ...

E. g. The main benefits of fiber are its exceptionally... Another benefit of fibers is that even when run along...

The main benefits of _____ Another benefit of _____.

2. ... in spans of ...

E. g. Optical cables can typically be installed in duct systems in spans of 6,000 m or more.

_____ in spans of _____.

3. ... ease of ...

E. g. Electrical transmission is often preferred because of its ease of operating transducers in linear mode.

_____ ease of _____.

4. ... has the capability to ...

E. g. Electrical transmission has the capability to carry electrical power as well as signals (in specially-designed cables).

_____ has the capability to _____.

5. ...in contrast to...

E.g. Fiber cables experience effectively no crosstalk, in contrast to some types of electrical transmission lines.

_____ in contrast to _____.

CAREER SPEAKING

Business Meeting

Ⅰ. Warm-Up

Model

(A: chairman　　B: Nancy　　C: Ben　　D: Ann)

A: Thank you for coming, everyone. It's really a pleasure to see you all here. Our main aim today is to discuss the time and mode of shipment. Would you like to make any comment on that?

B: Our customers hope we can ship the goods at once in order that they will catch up with the **season**.

C: We haven't received their **L/C**. According to the last **negotiation**, we won't arrange the shipment until the L/C arrives. Nancy, you'd better make sure whether they've sent the L/C.

B: OK, I take your point. They didn't mention the mode of shipment. I think we should produce the possible solution first so as to negotiate with them next week.

C: Then, do we have the goods sent by sea, by air or by railway?

B: I would prefer to have goods carried by railway. It's simple and cheap.

A: That sounds reasonable. What do you think about this proposal, Ann?

D: Well, why don't we choose sea shipment? It's safer.

B: Sometimes there is no direct **vessel**, and it's easy to cause a delay.

C: That's exactly the way I feel. How about by air? Since our customers worry about the time of delivery, it's the most convenient to ship by air.

D: I don't agree unless they pay more **freight**.

A: It seems a complex matter. We'll stop here for today and leave that to another time. Can we set the date for the next meeting, please?

Notes

season	n. 旺季
L/C	信用证(letter of credit)
negotiation	n. 谈判
vessel	n. 船
freight	n. 运费

II. Matching

Directions: Please choose the corresponding sentences from the box to fill in the blanks. Then play roles with your partners.

(A: Buddy   B: Henry   C: Christina   D: Wendy)

A: Good morning, everyone. If we are all here, let's get started. _____

B: In my opinion, it's necessary to redesign our office layout.

C: Right. _____

D: I also think so. The office can be decorated with green plants.

A: _____

D: Frankly speaking, most of them are old-fashioned and out of order, and this would lower the efficiency of work. _____

C: Unfortunately, I see it differently. It is a waste of money. You know, we have not enough funds.

D: _____

S1: What do you think of present office facilities?
S2: Why don't we update or replace them?
S3: Our main aim today is to study how to improve the working condition.
S4: But, at least the problem will be solved step by step.
S5: We need new environment for a change to get refreshed.

III. Closed Conversation

Directions: Please fill in the blanks according to the Chinese, then play roles with your partners.

(A: chairman   B: Lisa   C: Frank   D: Paul   E: Molly)

A: _____

(首先我欢迎你们并且感谢你们的到来,尤其是一通知你们就来了。我知道你们都很忙,从日常工作中抽出时间来参加这次会议是不容易的。)We're here today to discuss the matter of packing. Any suggestions on packing are greatly appreciated.

B: _____
(在我看来,漂亮的包装有助于推销产品,加深消费者对我们品牌的印象。)So our packing should be exquisite in design.

D: I agree. Consumers are easy to be attracted by beautiful appearance.

A: _____(你是怎么想的,弗兰克?)

C: I think the key point of packing lies in protecting the goods from moisture and shock.

D: _____(用木箱怎么样?)

B: The cost will be much higher. _____(我更青睐纸箱。) They are cheaper, lighter and easier to handle.

C: That's quite true, but I'm afraid the cardboard boxes are not strong enough for transportation.

E: May I have a word?

A: Of course, Molly. _____(请继续。)

E: I suggest foam plastics and an inner waterproof lining should be applied to protect the goods. In addition, we can reinforce the packing with wooden straps.

A: _____
(谢谢大家的好主意,我们在下周的会议上再继续讨论。)

Ⅳ. Semi-Open Conversation

Directions: Please fill in the blanks with your own words, then play roles with your partners.

(A: chairman   B: Mary   C: Sam   D: May)

A: Let's get this meeting started. Is everyone here?

B: I don't think Peter is here.

A: Okay. We can't wait. _____ First please explain briefly what your departments accomplished during the three weeks. Sam, have you finished the financial report yet?

C: Yes. _____

A: Good. How about marketing department, Mary?

B: We have been developing our new marketing campaign.

A: _____

B: Yes. We've opened up a new outlet.

A: Well done! May, you are the head of Customer Service Department. _____

D: My department has been doing a survey and collecting all the necessary data, but some customers refuse to cooperate.

A: Why?

D: I'm not sure. Sue is looking into that.

A: _____

E: Yes, I will.

A: Good luck to you. _____

V. Live Show

Challenge 1

Directions: Lucy holds a business meeting. Betty, Susan, Roy and Martin attend it. They discuss the subject of packing and shipping of products. Please act it out with your partner or partners. The model and the useful expressions below are just for your reference.

Challenge 2

Directions: Please hold a meeting to discuss the arrangement of receiving American guests. Act out in groups. The model and the useful expressions below are just for your reference.

Model

(A: Ann   B: Bob   C: Clair   D: David   E: Emily)

A: Good morning, everyone. Please be seated and we'll get started. As you can see here on the **agenda** we will be mainly talking about three issues: one, the design of a new advertising **brochure**; two, the approaches to cost reduction; and, three, the improvement of the customer service. Now, let's begin with the first issue. Emily, what's your opinion of the designs of our present advertising brochure?

E: I'm fed up with the old brochure. The information's out of date, and the telephone numbers are all wrong.

A: Right. So would it help to think what we want to put in the brochure?

C: It should give a brief introduction to the new product and indicate its quality and features. A few photographs and an approach to contact are added to it. Simple as that.

D: I was thinking that the brochure should be a bit more **"visual"**. We can use lots of photos of the product in use, the manufacturing process, satisfied customers and something else. More data and fewer words will have a better effect. What do you think?

E: I'm with you. In order to guide our **campaigns** to success in the new market, the designs of the brochure must be **novel** and **appealing** to eyes.

A: Could you put it in detail, Emily?

E: I am not an expert on designing. In my view, we can contact a design agency and get it to use our rough ideas to come up with a few **alternatives**.

B: That will cost a lot of money, I suppose.

E: But for a new company, such expenses are absolutely necessary.

D: I follow you, Emily. It's worth doing as long as the result is satisfactory.

A: I never thought of that. Maybe it works. Clair, what do you think about this proposal?

C: That's a good idea. We can have a try.

A: Well, David, would you like to contact several design agencies and get **quotations**?

D: Yes, no problem.

A: Bob and Clair, you two are in charge of the text for the brochure. OK?

B and C: OK.

Notes

agenda *n.* 议事日程　　　　　　　brochure *n.* 宣传手册

visual *adj.* 直观的，视觉的　　　campaign *n.* 广告计划

novel *adj.* 新颖的　　　　　　　appealing *adj.* 吸引人的

alternative *n.* 选择　　　　　　　quotation *n.* 估价

Useful expressions

1. 开始会议

Let's begin.

Let's get down to the business.

Shall we begin?

2. 表达意见

I (really) feel that ...

In my opinion ...

If you ask me, I tend to think that ...

3. 表示同意

I totally agree with you.

That's (exactly) the way I feel.

I couldn't agree more.

4. 表示异议

Up to a point I agree with you, but ...

I'm sorry, but I'm not sure I agree.

(I'm afraid) I can't agree.

5. 要求发言

May I have a word?

Can I say something here?

Excuse me for interrupting.

6. 询问意见

Can we get your view?

How do you feel about ... ?

Has anyone else got anything to contribute?

7. 结束会议
We'll stop here for today.
Let's call it a day.
If there are no other issues to discuss, I'd like to wrap up this meeting.

WRITING

Memo

备忘录(memo，memorandum 的简称)是以书面形式来传递内部各种事务信息的一种方式,可以起到通知变化、提醒事项、记录电话或会议内容等作用。备忘录结构明了,语言简洁,方便阅读与传递。其格式一般较为固定,通常分为开头和正文两部分。开头包括 To:(收信人姓名、职务)、From:(发信人或留言人姓名、职务)、Date:(日期)和 Subject:(主题)。其格式如下。

```
                          Memo
To:
From:
Date:
Subject:
```

在写备忘录时要注意:表达清晰,结构紧凑;开门见山,无须寒暄;一般不采用信头称呼的方式。

Sample 1

Memo

To: John Henry, Sales Manager
From: Sue Clinton, Saleswoman from ABC Foreign Trade Company
Date: June 8th, 2011
Subject: Shipment of the Scanners
I'm calling about the shipment of the scanners coming from Nanjing. Our customers

are badly in need of the goods. Please call me back at 695-8942 before this Friday.

Sample 2

Memo

To: All Staff

From: Jane Jackson, Marketing Manager

Date: May 30th, 2011

Subject: Meeting on Marketing and Promotion

We'll have a meeting in the conference room at 10:00a.m. on June 5th. The purpose of the meeting is to discuss how to increase market shares and how to promote new products. Any suggestions and ideas are welcome.

Useful expressions

1. I have several proposals for ...
2. We'll discuss ... at the meeting.
3. Please let me know your reactions to the plans.
4. Here we announce the appointment of ...
5. I have the pleasure of announcing that ...

Challenge

On the morning of February 2nd, 2011, Wang Lili, a clerk from Wuhan Import & Export Company, calls Fang Xin, marketing manager of your company, but he is not available. Wang Lili would like to make an appointment with Fang Xin to talk about the details of shipment. She will call him again this Thursday. Please write a memo.

EXTENSIVE READING

6.5 Optical Fibers

An optical fiber consists of a core, **cladding**, and a **buffer**(a protective outer coating), in which the cladding guides the light along the core by using the method of total internal reflection. The core and the cladding (which has a lower-refractive-index) are usually made of high-quality **silica** glass, although they can both be made of plastic as well. Connecting two optical fibers is done by **fusion** splicing or mechanical splicing and requires special skills and interconnection technology due to the microscopic precision required to align the fiber cores.

Two main types of optical fibers used in optic communications include multi-mode optical fibers and single-mode optical fibers. A multi-mode optical fiber has a larger core (≥ 50 mm), allowing less precise, cheaper transmitters and receivers to connect to it as well as cheaper connectors. However, a multi-mode fiber introduces multimode distortion, which often limits the bandwidth and length of the link. Furthermore, because of its higher **dopant** content, multi-mode fibers are usually expensive and exhibit higher attenuation. The core of a single-mode fiber is smaller (<10 mm) and requires more expensive components and interconnection methods, but allows much longer, higher-performance links. In order to package fiber into a commercially-viable product, it is typically protectively-coated by using ultraviolet (UV), light-cured **acrylate polymers**, then terminated with optical fiber connectors, and finally assembled into a cable.

After that, it can be laid in the ground and then run through the walls of a building and deployed aerially in a manner similar to copper cables. These fibers require less maintenance than common twisted pair wires, once they are deployed.

Notes

cladding *n.* 包层
silica *n.* 石英
dopant *n.* 掺杂物
buffer *n.* 保护层
fusion *n.* 融化
acrylate polymer 丙烯酸酯聚合物

Exercise

Directions: Match the two columns.

1. less signal degradation
2. non-flammable
3. digital signal
4. higher carrying capacity

A. useful in computer networks
B. more phone lines to go over the same cable or more channels to come through
C. transmit small signals
D. transmit in heat condition

6.6 Fiber to the Home

Fiber to the Home or FTTH is a telecommunications network based on optical fiber. Traditionally most networks were based on a copper or coaxial cables. Optical fiber networks allow for a high **volume** of data transfer compared to other networks. This includes TV, internet and phone services (so called triple play).

Over the last years companies like Verizon and AT&T have deployed fiber networks across the USA. Fiber to the Home, or FTTH, means the fiber runs all the way to your home or apartment, vs e.g. FTTC or Fiber to the Curb where the fiber runs till your curb and the existing copper infrastructure is used for the last yards of data transfer (and hence forming the bottle neck).

Optical Fiber has been used since a long time for long haul communications (e.g. the cables lying on the bottom of our ocean which take care of communications between the continents), however FTTH is a rather new development with high expectations. Many countries and areas in Asia are leading in the access of FTTH by customers. Leading countries and areas are South Korea, Japan and Hong Kong. In Europe especially the Nordic countries are leading. In the US several companies are deploying a fiber to the home, FTTH, and have started with the most populated areas. The Obama government stimulus package contains incentives for the deployment of FTTH in rural areas. These areas are often served by small rural telecommunications companies. South Korea, in particular, is a world leader with more than 31 percent of its households boasting FTTH broadband connections. Hong Kong is second globally, with more than 23 percent penetration, while Japan is a close third with more than 21 percent of its households FTTH ready.

Although fiber-optic systems excel in high-bandwidth applications, optical fiber has been slow to achieve its goal of "Fiber To The Home" or to solve the last mile problem.

Exercise

Directions: Answer the questions.

1. What is triple play?
_____
_____

2. Which country is the world leader in FTTH market penetration?
_____
_____

3. Which year is the best year in terms of new FTTH subscribers worldwide?
_____
_____

TRANSLATION

Translation of Adverbial Clauses

状语从句的翻译

英语的状语从句一般由从属连接词和起连词作用的词组来引导。英语的状语从句包括时间、目的、地点、原因、条件、让步、结果、比较等从句,可采用顺译法、换序法、分译法、合译法、转换法,以及几种方法同时采用的并用法。

1. 顺译法

把状语从句直接汉译,并与原文位置相同。

This process continues **till an equilibrium is established.**

这个过程持续下去**直到达到平衡**。

2. 换序法

依据汉语语言习惯或表意需要,状语从句汉译后与原文相比,位置发生变化,或前移,或后移,或置中。

Usually the capacitor is made up of plates of large area **so that large electrical charges may be stored.**

为了能储存大量电荷,电容器通常用大面积的金属板来制造。

3. 分译法

有的状语从句非常长，且结构复杂。采用分译法，能清楚表达长句的意思。

Albert Einstein's paper — *On the Quantum Theory of Radiation* in 1917 was considered as a great event, **because it laid the foundation for the invention of the laser and its predecessor, the maser, in a ground-breaking rederivation of Max Planck's law of radiation based on the concepts of probability coefficients（later to be termed "Einstein coefficients"）for the absorption, spontaneous and stimulated emission.**

阿尔伯特·爱因斯坦的论文《论辐射的量子性》在1917年发表被认为是重大事件，因为该论文为激光器及其先驱——微波激射器的发明奠定了基础。这项成果源于对马克斯·普朗克辐射定律创造性的再次运用，而马克斯·普朗克辐射定律基于概率系数理论（该系数后被称为"爱因斯坦系数"），用于研究受激吸收、自发辐射和受激辐射。

4. 合译法

含有状语从句的复合句在汉译时可省去主句的主语或从句的主语，将两个句子组合成一个单句。

The iron should be stricken while it is hot.

趁热打铁。

5. 转换法

译文根据需要，将一类状语从句译为另一类从句或词组，如条件状语从句译为时间状语从句。

If all the steps are taken in accordance with the procedures, the system will be recovered.

当依程序采取所有步骤时，系统将恢复。

（转译为时间状语从句）

6. 并用法

由于表意需要，有些状语从句在汉译时需要以上几种方法并用。

Shall the atoms influence process of change if the atoms move?

原子运动是否会影响变化的进程？

译文采用换序法，调整主句与状语从句的语序，将状语从句提前，主句放后；同时，采用了合译法，将主句和从句合成一个单句。

Challenge

Translate the following sentences into Chinese.

1. Whenever a wave moves out from a source in uniform medium, it travels in straight lines.
2. Where there is the Internet, there is chatting.
3. The path is not completed till wires are connected.
4. Current stops flowing as soon as we break the circuit.
5. Chemical reaction occurs if possible.
6. Information becomes jammed in a bottleneck at the points where conversion to, or

from, electronic signals is taking place.

7. As the sphere becomes larger, the waves become weaker.

TASK

Fiber-Optic Communication

Target	Understand and express professional knowledge.
Requirements	Make an English presentation according to Unit 6. Produce a micro-video to express professional knowledge. Add accurate Chinese and English subtitles to the video.
Contents	Fiber-optic communication
Group: _____		Project	Name	Software	Time	Score
		Material collection				
		Script				
		PPT				
		Reader				
		Subtitle				
		MP4				
Operating environment	XP/Win7/Win8/ Mobile phone
Product features	

Script reference（略）

UNIQUE ENGLISH

Mother

Written by Howard Johnson

"M" is for the million things she gave me,

"O" means only that she's growing old,
"T" is for the tears she shed to save me,
"H" is for her heart of the purest gold,
"E" is for her eyes, with love-light shining,
"R" means right, and right she'll always be,
Put them all together, they spell "MOTHER",
A word that means the world to me.

References

[1] Agrawal G P. Fiber-Optic Communication Systems[M]. Hoboken：John Wiley & Sons，Inc. ，2002.

[2] Osche G R. Optical Detection Theory for Laser Applications[M]. Hoboken：John Wiley & Sons，Inc. ，2002.

[3] Steen W M，Mazumder J. Laser Material Processing [M]. 4th ed. London：Springer，2010.

[4] Lawrence J . Advances in Laser Materials Processing ：Technology，Research and Applications [M]. Oxford ：Woodhead Publishing Limited，2010.

[5] Laser Institute of Ameica. Laser Safety Information Bulletin [E] [2010-9-11].

[6] TLF Laser in Material Machining. Laser Welding [E] [2011-2-9]. Laser Material Processing [E] [2010-12-16].

[7] Koechner W. Solid-State Laser Engineering [M]. 5th ed. 北京：世界图书出版公司，2005.

[8] [美]约瑟夫·C·帕勒里斯. 光纤通信[M]. 北京：电子工业出版社，2005.

[9] 王琳，夏怡. 电子与通信专业英语[M]. 北京：北京理工大学出版社，2007.

[10] 刘小芹，刘骋. 电子与通信技术专业英语[M]. 2版. 北京：人民邮电出版社，2008.

[11] 李霞. 电子与通信专业英语[M]. 北京：电子工业出版社，2005.

[12] 刘其斌. 激光加工技术及其应用[M]. 北京：冶金工业出版社，2007.

[13] 教育部. 教育部关于全面提高高等职业教育教学质量的若干意见[Z]. 教高〔2006〕16号.

[14] 王九程，袁智英，黄丛笑. 高职《公共英语》的职业性及其提高探讨[J]. 湖北广播电视大学学报，2010，(11)：37-38.

[15] 刘文宇，王慧莉，张旭. 商务英语口语大全[M]. 大连：大连理工大学出版社，2007.

[16] 余啸海，薛昆. 实用英语口语速查手册[M]. 北京：国防工业出版社，2005.

[17] 蒲村，刘华，燕静敏，等. 英语会话全程通[M]. 北京：机械工业出版社，2003.

[18] 卢欣，王晓妍. 1+X脱口说：商务英语口语[M]. 大连：大连理工大学出版社，2007.

[19] 张翠萍. 商贸英语口语大全[M]. 北京：对外经济贸易大学出版社，2006.

[20] 盛小利. 商务英语谈判口语[M]. 北京：中国宇航出版社，2007.

[21] 刘一平，李宏亮. 商务英语口语[M]. 北京：北京大学出版社，2006.

[22] 教育部《机电英语》教材编写组. 机电英语[M]. 北京：高等教育出版社，2001.

[23] 赵萱，郑仰成. 科技英语翻译[M]. 北京：外语教学与研究出版社，2006.

[24] 张朴顺. 现代服务业英语会话900句[M]. 北京：中国标准出版社，2005.

[25] 谢江南，何加红. 实用英文写作[M]. 2版. 北京：首都经济贸易大学出版社，2005.

[26] 张道真.实用英语语法(1995年修订重印本)[M].北京:外语教学与研究出版社,1995.
[27] 陈莉萍.专门用途英语存在的依据[J].外语与外语教学,2001(12):28-30.
[28] 范谊.ESP存在的理据[J].外语教学与研究,1995(3):43-48.
[29] 冯世梅,杜耀文.现代科技英语词汇与翻译[J].中国科技翻译,2002,15(4):52-54.
[30] 冯志杰.汉英科技翻译指要[M].北京:中国对外翻译出版公司,2001.
[31] 郭海平.科技英语词汇的构词特点及翻译[J].武汉轻工大学学报,2004,23(2):115-117.
[32] 刘双峰,王高.光机电系统概论[M].北京:北京理工大学出版社,2007.
[33] 克希奈尔.固体激光工程[M].孙文,江泽文,程国祥,译.北京:科学出版社,2002.
[34] 卫乃兴,周俊英.也谈ESP与大学英语教学[J].外语界,1994(2):32-36.
[35] 张少雄."科技英语词汇"评说[J].外语教学与研究,1994(3):47-50.
[36] 赵萱,郑仰成.科技英语翻译[M].北京:外语教学与研究出版社,2006.
[37] 朱一纶.电子技术专业英语[M].北京:电子工业出版社,2006.